THE SILENT EXPLOSION

Reactions That Release Massive Energy Without Sound or Flame

Anton Kesting

This book is a work of nonfiction. The information contained in this book is based on research, scientific literature, and publicly available reports and is presented for general informational and educational purposes only. While every effort has been made to ensure the accuracy of the information presented, the author makes no warranties, express or implied, regarding the completeness or accuracy of the contents, and assumes no liability for any errors or omissions, or for any actions taken in reliance on the information provided herein.

This book does not provide medical, safety, engineering, or legal advice. Readers facing specific technical, safety, or health concerns related to any topic discussed in this book should consult a qualified professional.

CONTENTS

INTRODUCTION

A Word Before We Begin

You are holding a book about explosions that make no sound.

That sentence should feel like a contradiction. For most of human history, the word explosion has meant one thing: a violent, deafening release of energy, accompanied by fire, smoke, and a shockwave you can feel in your chest. Dynamite. Gunpowder. A car backfiring. A star going supernova. When we imagine an explosion, we imagine noise.

But noise is not what an explosion actually is. Noise is a side effect. It is the pressure wave that happens to be produced when energy escapes a system fast enough to compress the air around it into an audible thump. Strip away the sound, strip away the flame, and what remains is the real definition of an explosion: a sudden, dramatic release of stored energy. Once you accept that definition, the entire concept opens up. Because it turns out that some of the most powerful energy releases on Earth, and off it, happen without a single decibel of noise, without a spark, without a flame you could warm your hands over.

A block of rusting steel, sitting quietly in a scrapyard, is technically undergoing combustion. It is burning, in the strict chemical sense, releasing the same category of energy that a campfire releases. It just does it so slowly that you would need a very sensitive thermometer to notice.

A chunk of plutonium the size of a golf ball, sealed inside a spacecraft bound for the edge of the solar system, radiates enough heat to warm a small room, continuously, for decades, without burning anything at all. No oxygen required. No flame possible. Just atoms quietly falling apart.

A pile of damp hay, stacked improperly in a barn, can reach temperatures above four hundred degrees Fahrenheit and burst into flame from the inside out, with no match, no wiring fault, no external heat source of any kind. The fire that eventually appears is not the beginning of the story. It is the end of a silent chemical reaction that had been building for days.

A lithium battery in your pocket, mostly by design, contains something close to an explosive amount of stored chemical energy in an extremely compact space. Under normal conditions it releases that energy in a slow, controlled trickle, a few watts at a time, so gently that you barely notice the warmth. Under abnormal conditions, that same battery can release its entire stored energy in seconds, and the results have grounded aircraft, triggered a fifteen billion dollar corporate recall, and forced airlines worldwide to rewrite their safety protocols.

This is a book about that hidden category of event. The silent explosion. It sits in the space between two things we think we understand well, energy and noise, and it reveals that we conflated them by accident. Sound is not a requirement of power. Flame is not a requirement of destruction, or of usefulness, or of wonder. Some of the most consequential reactions on this planet, the ones that built dams, warmed distant spacecraft, decomposed the bodies of extinct animals into the oil we now burn, and occasionally destroy entire buildings from within, do so in complete silence, at least until something goes wrong.

I wrote this book because I kept running into the same experience, over and over, while researching seemingly unrelated topics. I would be reading about nuclear power and stumble onto a fact about radioactive decay heat that seemed to belong in a different book entirely, one about volcanoes, or fires, or corrosion. I would be reading about lithium battery safety and find myself, unexpectedly, thinking about compost heaps. The connections kept multiplying until I realized they were not connections at all. They were the same phenomenon wearing different costumes: chemical and physical systems releasing stored energy without the sensory signature we have been trained since childhood to associate with power.

Every chapter in this book stands on real, verifiable science. Where a number appears, it comes from a measurement, a government report, a peer reviewed paper, or a documented historical event. Where a historical incident is described, it is a real incident, with real consequences, drawn from public investigation records and reputable reporting. I have made no attempt to invent drama where the facts did not already provide it, because in every single case I researched, the facts were dramatic enough on their own.

You do not need a background in chemistry, physics, or engineering to read this book. You need curiosity, and perhaps a willingness to look at ordinary things, a rusting nail, a bag of quicklime, a phone charging on your nightstand, and see them differently afterward. That is the only outcome I am hoping for. Not that you memorize a set of facts, but that the next time you notice something quietly generating heat with no visible cause, you pause, and you understand exactly what is happening inside it.

Let us begin where all explosions begin, loud or silent: with energy that has nowhere left to hide.

Anton Kesting

CHAPTER ONE

The Myth of the Bang

In October 2017, engineers at the Cassini spacecraft mission center watched their instrument readings go quiet as the probe descended into Saturn's atmosphere and was torn apart by heat and pressure. The mission had lasted nearly twenty years. For most of that time, Cassini's onboard power came not from solar panels, which are nearly useless that far from the sun, but from three small canisters packed with plutonium dioxide. Those canisters had been quietly converting radioactive decay into electricity since 1997, without a single moving part, without combustion, without noise. Thirty three kilograms of plutonium oxide produced roughly eight hundred and seventy watts of continuous power for two decades, silently, in the vacuum of space. By any reasonable definition, that plutonium was releasing an explosive quantity of cumulative energy. It just did so a few hundred watts at a time, spread across twenty years, rather than all at once in a fireball.

Compare that to what most people picture when they hear the word explosion. A demolition charge bringing down a building. A grenade. The Hindenburg. These images share three features: they are fast, they are loud, and they are visually spectacular. We have built our entire cultural vocabulary of power around these three features, to the point that we have started to believe they are inseparable from the concept of a large energy release. They are not. They are simply the most common way that energy release manifests when it happens over a very short timescale in the presence of a compressible gas, meaning air.

This chapter exists to take that assumption apart, carefully, because the rest of this book depends on you seeing past it.

What An Explosion Actually Is

Strip the definition down to its physics and an explosion is simply this: a rapid transition of a system from a higher energy state to a lower one, releasing the difference as heat, light, kinetic energy, or radiation. Nothing in that definition requires sound. Nothing requires flame. The definition requires speed, and even that requirement is more about human perception than about physics.

Consider what actually produces the bang in a conventional explosion. When dynamite detonates, the chemical reaction converts solid nitroglycerin into a mixture of hot gases, carbon dioxide, water vapor, nitrogen, and oxygen, in a fraction of a millisecond. Those gases, suddenly existing at extremely high temperature, try to occupy far more volume than the solid explosive did. The surrounding air gets shoved outward violently, faster than the speed of sound in some cases, creating a shockwave. That shockwave, when it reaches your ear, is what you perceive as the bang. The sound is not the explosion. The sound is the explosion's exhaust note, produced because the reaction happened to occur in an atmosphere and happened to produce gaseous products at extreme speed.

Now take that same amount of chemical energy and release it slowly instead. Say, over the course of a year rather than a millisecond. The same number of joules escape the system. The same bonds break, the same electrons rearrange, the same fundamental physics operates. But there is no shockwave, because nothing is moving fast enough to compress the air violently. There is no flame, if the reaction does not require open combustion in oxygen. There is just heat, quietly leaving a system, watt by watt, joule by joule, until the stored energy is gone.

This is not a hypothetical. It is precisely what happens inside a block of plutonium 238, the isotope used in those RTGs, or radioisotope thermoelectric generators, that powered Cassini, and Voyager 1 and 2, and the Curiosity and Perseverance rovers on Mars. Plutonium 238 has a half life of 87.7 years, decaying primarily through alpha emission, and a pure sample generates approximately 0.54 to 0.57 watts of thermal power per gram, continuously, for decades. A sphere of the stuff about the size of a grapefruit will sit there, radiating heat you could feel from several feet away, hot enough that early handling required tongs and shielding, and it will never once make a sound or produce a visible flame. It cannot. There is no chemical combustion happening, no rapid gas expansion, no oxygen involved at all. The energy comes from the nucleus of the atom itself, slowly shedding mass as it transforms into a more stable configuration, and it escapes as pure thermal radiation.

If you added up the total energy that a kilogram of plutonium 238 will release over its full decay lifetime, you would find a number that rivals a serious conventional explosive by weight. The difference is entirely in the release rate. Dynamite tries to dump its energy in under a millisecond. Plutonium 238 spreads the same category of energy release across roughly a century. Same fundamental category of event, wildly different timescale, and therefore a completely different sensory experience for anyone standing nearby.

Why Our Brains Default to Bang

There is a reason humans built an entire vocabulary around loud, visible energy release, and it has nothing to do with physics and everything to do with survival. For nearly all of human evolutionary history, the energy releases that mattered to your survival were fast ones. A predator's strike. A falling rock. A wildfire crowning through a forest canopy. These events shared the qualities of speed, noise, and visual violence, and organisms that reacted strongly to those cues tended to survive better than organisms that did not. Our nervous systems became finely tuned danger detectors calibrated specifically for fast, loud, visible threats.

Slow energy release, by contrast, posed little direct threat to a hunter gatherer. A rusting rock was not going to attack you. A radioactive mineral deposit, unless you happened to be a Curie a hundred and thirty years before its discovery, produced no sensory signal a human nervous system was built to detect. So we never developed an instinctive fear of, or fascination with, slow energy release, even when the total quantity of energy involved dwarfs what a bomb produces. Our attention systems are simply not wired to flag it as significant.

This has real consequences beyond simple curiosity. It means that as a species, we have historically underestimated the danger, and the usefulness, of slow energy release phenomena, because they do not trigger our alarm systems the way a bang does. Compost heaps have burned down barns for centuries before anyone understood why, precisely because nobody could point to a spark, a flame source, or a loud event that preceded the fire. The fire itself was mistaken for the beginning of the danger, when in fact the danger had been building silently for days inside the pile, hidden entirely from human senses until the moment it crossed a threshold and became visible.

The same blind spot shows up in modern engineering failures. Lithium battery fires, which we will examine closely in a later chapter, frequently catch people off guard specifically because the battery gave no loud warning before failing. A device sitting quietly on a nightstand, charging normally, can undergo an internal short circuit that triggers what battery engineers call thermal runaway, a self accelerating chain reaction where heat generation outpaces heat dissipation. Depending on the cell chemistry and the failure mode, this process can unfold over many seconds or even minutes before it becomes visible or audible, and during that entire silent buildup phase, the battery gives almost no external sign that anything is wrong. By the time you hear a hiss or see smoke, the reaction is often already well past the point where it could be stopped.

A Different Way to Measure Power

If sound and flame are unreliable indicators of how much energy a system is releasing, what should we use instead? Physicists and engineers rely on a small set of measurements that strip away the sensory drama and get straight to the substance.

The first is simply total energy, usually measured in joules, or for larger quantities, kilojoules or megajoules. This tells you how much energy a system releases in total, regardless of how fast it happens. Combustion of one mole of methane gas, the primary component of natural gas, releases

890.4 kilojoules of energy. That number does not change whether the methane burns in a campfire over several minutes or detonates in a mining accident in under a second. The total energy is the total energy. What changes between those two scenarios is the second measurement that matters: power, which is energy divided by time, usually measured in watts, where one watt equals one joule per second.

This distinction, energy versus power, is the single most important concept in this entire book, and it is worth sitting with for a moment. Energy tells you how much fuel is in the tank. Power tells you how fast you are burning through it. A city bus and a sports car might have the exact same size fuel tank, holding the exact same total chemical energy, but the sports car can release that energy far faster, producing far more power at any given instant, which is why it accelerates harder and why, if something went wrong with its fuel system, the failure would likely be more violent.

Apply this framework to plutonium 238 and rusting steel and you start to see the pattern clearly. Rusting steel and burning steel release the exact same category of chemical energy, oxidation of iron, and can in principle release a very similar total quantity of energy per kilogram of iron consumed. The reaction 4Fe plus 3O2 yields 2Fe2O3, and releases about 1829 kilojoules of energy for every four moles of iron oxidized, using standard thermochemical values. What differs enormously between rust and fire is power, the rate at which that fixed quantity of energy gets released. A rusting nail might release that energy over months or years, at a power output so low it is genuinely difficult to detect with anything short of specialized laboratory equipment. A burning piece of steel wool, ignited by a spark, releases a similar quantity of energy per gram in seconds, producing enough power to visibly glow and, under the right conditions, sustain a flame.

Same chemistry. Same fundamental reaction. Radically different power output, and therefore a radically different sensory experience.

The Explosions History Forgot to Name

Because our language evolved around fast, loud events, we never developed a good vocabulary for slow, silent ones, even when they matter enormously. We do not call a rusting bridge support an explosion, even though it is releasing chemical energy through the same basic oxidation process that powers a bonfire. We do not call the curing of a massive concrete structure an explosion, even though the chemical reaction involved, the hydration of Portland cement, releases genuine heat, sometimes enough heat to threaten the structural integrity of the concrete itself if it is not carefully managed.

This gap in vocabulary is not merely a linguistic curiosity. It has practical consequences for how seriously we take slow energy release phenomena, and how well we engineer around them.

Take the Hoover Dam as an example, which we will explore in more depth in a later chapter. When construction began in 1931, engineers faced an unusual problem. The dam required over three million cubic yards of concrete, and concrete, as it cures, generates heat through the hydration reaction between cement and water. In a small concrete slab, this heat dissipates harmlessly into the surrounding air. In a mass as enormous as the core of Hoover Dam, the heat had nowhere to go. Engineers calculated that if the dam had been left to cure naturally, without intervention, the concrete at its core would have taken roughly one hundred and twenty five years to fully cool. Long before that cooling completed, the temperature differential between the hot interior and the cooler exterior would have caused the concrete to crack catastrophically, compromising the entire structure.

Nobody calls the curing of Hoover Dam's concrete an explosion. There was no bang, no flame, no visible violence. And yet the engineers who built it understood, correctly, that they were managing an enormous, dangerous, silent energy release, every bit as real as a chemical explosion, just spread across a timescale of years instead of milliseconds. They solved the problem by embedding nearly six hundred miles of steel cooling pipes throughout the concrete, circulating ice water continuously through the curing structure to extract heat faster than the natural rate would have allowed. Ice for that system was produced at a rate of over a thousand pounds per day at a dedicated refrigeration plant built specifically for the dam project. This was, in every meaningful sense, an engineering effort to control an explosion. It simply never got called one, because it never made a sound.

Setting the Terms for This Book

For the remainder of this book, when the word explosion appears, understand it in its broadest, most physically accurate sense: any sudden or sustained release of stored energy from a system, regardless of whether that release produces sound, light, or flame. Some of the explosions in the chapters ahead will indeed be loud, because contrast is useful and because some silent energy release systems occasionally cross a threshold and become audible ones, as we saw with lithium battery fires. Most, however, will not make a sound at all, and that silence is precisely what makes them worth studying closely.

You will meet radioactive isotopes that have quietly powered spacecraft for half a century. You will meet compost piles that generate enough internal heat to spontaneously combust from within, no external ignition source required. You will meet batteries that store an almost explosive quantity of chemical energy in a space smaller than a deck of cards, releasing it gently under normal use and catastrophically under failure. You will meet rust, concrete, ice, and the human body itself, each one revealed as a site of genuine, measurable, sometimes enormous energy transformation happening entirely outside the range of what our senses are built to detect.

The myth of the bang is comfortable because it is simple. Loud and visible equals powerful. Quiet and invisible equals harmless. That myth has shaped how we build things, how we regulate hazards, and how we teach science to children who grow up thinking an explosion always looks like the ones in movies. It is also, provably, false, and the cost of believing it, as you will see in the chapters on batteries, silos, and barns, has occasionally been counted in lives and in billions of dollars of damage.

Once you let go of the myth, the world becomes considerably more interesting. Every quietly rusting object, every warm patch of soil, every battery humming faintly on a charger, becomes a small window into the same physics that powers a bomb, a star, or a spacecraft two hundred million kilometers from home. The energy was always there. We simply were not trained to notice it until it announced itself with noise.

The next chapter takes this idea and grounds it firmly in the physics that governs it: the laws of thermodynamics, the concept of activation energy, and the precise reasons why some reactions release their energy in a millisecond while others take a century. Understanding that difference is the key that unlocks everything that follows.

CHAPTER TWO

The Physics of Silence

Why does one match light a campfire in seconds while a buried log takes centuries to decompose into peat, releasing roughly the same category of chemical energy along the way? Why does gasoline vapor detonate with a shockwave while the slow oxidation of the same hydrocarbons inside your body, during ordinary metabolism, produces no shockwave at all? The answer lives inside two ideas that most people encounter briefly in a school chemistry class and then never think about again: activation energy and reaction rate. Understood properly, these two ideas explain almost everything in this book.

The Energy Hill

Imagine every chemical reaction as a ball sitting at the top of a hill, ready to roll down into a valley. The valley represents a lower energy state, more stable, more settled. The ball rolling down releases energy as it goes, the same way a boulder rolling downhill releases kinetic energy. That released energy is what we call exothermic, meaning heat leaving the system.

But here is the detail that changes everything: the ball does not simply roll down the hill on its own. First, it usually needs to be pushed up a small rise near the top, called the activation energy barrier, before gravity can take over and send it down the far side into the valley. This is true even for reactions that are hugely exothermic overall. Gasoline and oxygen, sitting together in an open container, do not spontaneously combust, even though the reaction between them releases enormous energy once it starts. They need a spark, a small input of energy, to push the reaction over that initial barrier. Once pushed over, the reaction proceeds on its own, releasing far more energy than the spark contributed, and that surplus energy helps push neighboring gasoline molecules over their own barriers too, creating the self sustaining chain reaction we recognize as a flame.

Activation energy explains why some hugely energetic reactions do not happen instantly, all the time, everywhere. Paper is chemically capable of releasing significant energy through combustion with atmospheric oxygen, and yet books on your shelf are not spontaneously bursting into flame, because the activation energy barrier for that reaction, at room temperature, is high enough that essentially nothing in your living room has enough energy to clear it. Add a lit match, providing a burst of localized energy that clears the barrier for a small patch of paper, and the reaction proceeds, releasing enough heat to clear the barrier for the next patch of paper, and the next, in a chain that we experience as fire spreading.

Reaction Rate: The Other Half of the Story

Activation energy explains whether a reaction can happen at all under given conditions. Reaction rate explains how fast it happens once it starts, and this second factor is where the real distinction between loud and silent explosions lives.

Reaction rate depends on several variables working together. Temperature matters enormously, because higher temperature means molecules are moving faster and colliding more often and more forcefully, increasing the chance that any given collision has enough energy to clear the activation barrier. This relationship is not linear. It is closer to exponential, described by a formula chemists call the Arrhenius equation, which means that relatively small increases in temperature can produce dramatic increases in reaction speed. This exponential relationship is precisely why thermal runaway in batteries becomes so dangerous: a small temperature rise accelerates the reaction, which produces more heat, which further accelerates the reaction, in a feedback loop that can go from barely noticeable to catastrophic in well under a minute.

Concentration and surface area matter too. A reaction between two substances happens faster when there is more of each substance available to react, and faster still when more of each substance is physically exposed to the other. This is why a log burns slowly while sawdust from the same log, with its vastly greater surface area exposed to oxygen, can burn explosively fast, and why finely divided metal or grain dust, with an enormous surface area relative to their mass, are capable of far more violent combustion than a solid block of the same material. We will return to this exact principle in the chapter on grain silo and industrial dust explosions, because it explains one of the more counterintuitive hazards in modern industry: substances that are essentially inert as solid blocks becoming genuinely explosive once ground into fine powder.

Catalysts change reaction rate without being consumed themselves, by providing an alternative pathway with a lower activation energy barrier. Enzymes inside your own body are catalysts, and without them, most of the biochemical reactions that keep you alive would proceed so slowly at body temperature that you would not survive on the timescale those reactions would naturally take. A reaction that might take years to occur without a catalyst can occur in milliseconds with the right one present, which is part of why biological systems, discussed in a later chapter, are capable of extraordinarily precise, extraordinarily fast energy management despite operating at a gentle ninety eight point six degrees Fahrenheit.

And then there is the factor most relevant to this entire book: whether the reaction produces gaseous products, and how quickly.

Gas Is the Real Source of Bang

Here is the physical mechanism that most people never learn explicitly, even though it explains the difference between every silent reaction in this book and every loud one. Sound, in the context of an explosion, comes almost entirely from rapid gas expansion. Solids and liquids are nearly incompressible. When a solid or liquid changes chemically, even releasing substantial energy, the surrounding matter does not need to move much to accommodate the change, so no meaningful pressure wave gets generated. Gases are the opposite. They are highly compressible and expand dramatically when heated, and a reaction that suddenly creates a large volume of hot gas from a small volume of solid or liquid reactant will push the surrounding air violently outward, producing exactly the shockwave your ear interprets as a bang.

This single fact explains an enormous amount. Rust does not bang because iron oxidizing into iron oxide is a solid to solid transformation. No gas is produced, so there is no rapid volume change to compress the surrounding air. Radioactive decay does not bang because alpha and beta particles, while carrying real kinetic energy, are not creating a large expanding volume of gas at the reaction site, they are simply flying outward from a nucleus. Dynamite bangs specifically because the reaction converts a dense solid into a large volume of hot gas within microseconds, and that gas has to shove the surrounding air out of the way at extreme speed to make room for itself.

Understanding this mechanism lets you predict, with reasonable accuracy, whether any given energetic reaction will be loud or silent, just by asking two questions. First, how much gas does the reaction produce relative to the volume of solid or liquid reactant consumed. Second, how quickly does that gas production happen. A reaction that produces a large volume of gas very quickly will be loud. A reaction that produces little or no gas, regardless of speed, will be silent. A reaction that produces significant gas but does so slowly, allowing the gas to escape and dissipate gradually rather than compressing the surrounding air suddenly, will also tend to be quiet, even if the total energy and gas volume released is substantial.

This last category deserves attention because it is easy to miss. Composting organic matter releases carbon dioxide and other gases continuously as microorganisms break down plant material, and the total volume of gas released over the life of a large compost pile can be considerable. But because that gas escapes gradually, over days and weeks, rather than in a sudden burst, there is no shockwave, no bang, nothing for a human ear to detect. The same total quantity of gas, released in a fraction of a second instead of a fraction of a month, would sound like an explosion. The chemistry barely changed. The timescale changed everything about how a human observer would perceive it.

Deflagration and Detonation, Explained Properly

Within the world of combustion science, there is a formal distinction between two modes of energy release that both technically qualify as explosions but produce very different experiences, and understanding this distinction sharpens the picture considerably.

Deflagration describes a combustion reaction where the flame front propagates through the unburned material at subsonic speed, slower than the speed of sound in that material. Most fires you encounter in daily life, campfires, house fires, the flame on a gas stove, are deflagrations. The reaction spreads because heat from the burning portion transfers to the adjacent unburned portion, raising it above its ignition temperature, which then burns and heats the next portion in turn, in a relay that moves at a comparatively modest pace.

Detonation describes a combustion or decomposition reaction where the reaction front propagates faster than the speed of sound in the surrounding material, driven not primarily by heat transfer but by a supersonic shockwave that compresses and ignites material ahead of it almost instantaneously. This is what happens inside high explosives like TNT or dynamite. The shockwave itself is doing the work of triggering the next layer of reaction, rather than waiting for heat to conduct through the material, which is why detonation happens so much faster and produces so much more violent pressure effects than deflagration.

Between these two extremes sits an enormous range of intermediate behaviors, and one of the most important, and most misunderstood, involves what happens when a slow deflagration transitions into a full detonation. This transition, sometimes abbreviated DDT for deflagration to detonation transition, is exactly what makes certain industrial accidents so much more dangerous than they initially appear. A grain dust fire, for example, might begin as a relatively contained deflagration inside one section of a grain elevator. But the pressure wave from that initial deflagration can kick additional dust into suspension in the air ahead of the flame front, creating a fresh, densely packed fuel and air mixture exactly where the flame is about to arrive. Under the right conditions, this feeding mechanism can accelerate the flame front until it crosses the threshold into detonation, turning what began as a contained, survivable fire into a structure destroying blast. This mechanism has been documented in numerous grain elevator explosion investigations conducted by the United States Chemical Safety Board and by agricultural safety researchers, and it explains why grain dust, something most people would never associate with explosive violence, has caused some of the deadliest industrial accidents in agricultural history. We examine specific documented cases in the chapter on industrial dust explosions.

Silence Is a Spectrum, Not a Switch

It would be tidy if reactions sorted cleanly into loud and silent categories, but the physics does not work that way. What actually exists is a continuous spectrum of energy release rates, stretching from radioactive decay processes that take place over billions of years, through corrosion processes measured in decades, through composting and curing processes measured in days or weeks, through fire measured in minutes, all the way to detonation measured in microseconds. Sound only appears once you cross into the fast end of that spectrum, and even then, the volume and character of the sound depends heavily on exactly how fast the gas production occurs and how much gas is involved.

This spectrum framing matters because it reveals something the loud versus silent framing obscures: every explosive event in this book, no matter how quiet, sits on the exact same physical continuum as a bomb. There is no separate category of physics governing silent energy release. It is the identical thermodynamics, the identical concept of activation energy, the identical relationship between reaction rate and temperature, simply operating at a different point on the timescale. This is why a chemist or physicist can look at a rusting bridge, a curing dam, a decaying radioactive isotope, and a stick of dynamite and correctly describe all four using the exact same fundamental vocabulary of enthalpy, activation energy, and reaction kinetics. The dynamite just happens to be doing its chemistry very, very fast.

What Determines Where a Reaction Sits on the Spectrum

A handful of physical properties determine where any given reaction falls along this continuum from silent to explosive, and worth naming them clearly here because they will recur throughout the rest of this book.

The height of the activation energy barrier is the single largest factor. High barriers mean the reaction proceeds slowly at ordinary temperatures because few molecular collisions carry enough energy to clear the barrier, which is why rust takes years rather than seconds even though the overall energy release, per kilogram of iron, is comparable to burning the same iron directly.

The physical state of the products matters just as much, as we established above. Reactions confined to solids and liquids, without significant gas production, essentially cannot produce a loud bang no matter how much total energy they release, because there is no compressible medium being displaced suddenly.

The availability of a chain reaction mechanism matters enormously. Reactions that produce a byproduct capable of triggering further reaction in neighboring material, heat in the case of fire, neutrons in the case of nuclear fission, free radicals in the case of certain battery failure modes, can accelerate dramatically once started, because each individual reaction event effectively recruits its neighbors into reacting too. This is the mechanism behind thermal runaway in batteries, chain reactions in nuclear fission, and flame propagation in ordinary fire, and it is the single most important reason why some reactions that start slowly can accelerate into something violent given the right conditions.

And finally, confinement matters. A reaction that releases gas into an open, unconfined space will simply push that gas outward relatively gently, producing a soft whoosh rather than a sharp bang, which is why a propane leak igniting in open air produces a very different sound and pressure signature than the same quantity of propane igniting inside a sealed room. Confinement forces pressure to build before the gas can escape, and it is that buildup and sudden release that produces the sharpest, loudest, most destructive pressure waves. This is precisely why battery fires inside a sealed phone casing, or a laptop chassis, or an aircraft cargo hold, tend to be more violent than the same battery chemistry failing in open air, a detail with direct relevance to the aviation safety regulations discussed later in this book.

Carrying This Framework Forward

You now have the core physical vocabulary needed to understand every chapter that follows:

activation energy, reaction rate, the role of gas production and confinement in generating sound, and the spectrum that stretches from geological timescales to microseconds. Every silent explosion examined in the rest of this book, rust, radioactive decay, composting, concrete curing, battery failure, phase transitions, metabolism, geological processes, will make sense within this same framework. The chemistry differs from chapter to chapter. The physics governing why some of these reactions are loud and some are silent does not.

What should be clear by now is that silence is not the absence of power. It is simply the absence of gas produced quickly enough, in a confined enough space, to compress the air into an audible shockwave. The energy is there regardless. Sometimes it built a spacecraft's power supply. Sometimes it is quietly consuming a steel beam from the inside. Sometimes, given the right trigger, it accelerates from silence into catastrophe in less time than it takes to read this sentence. The next chapter begins with the most familiar silent explosion of all, one that surrounds nearly every piece of iron and steel on the planet, and one that most people have never once considered to be a form of fire.

CHAPTER THREE

Rust: The Slowest Fire on Earth

The SS Central America sank in a hurricane off the coast of the Carolinas in September 1857, carrying roughly three tons of gold and hundreds of passengers to the ocean floor, more than seven thousand feet down. When marine archaeologists finally located the wreck in 1988, they found something that surprised even experienced deep sea researchers. The ship's iron hull and iron fittings had transformed into massive, fragile structures called rusticles, elaborate hollow formations of iron oxide that looked almost organic, hanging from the wreck like stalactites made of rust. The gold, being chemically inert, sat untouched exactly as it had for over a century. The iron, meanwhile, had been quietly burning the entire time, one molecule at a time, for a hundred and thirty one years, in water so deep and so cold that almost no light reaches it.

That word, burning, is not a metaphor. It is the technically correct term for what rust actually is.

The Chemistry Nobody Calls Fire

Combustion, in the strict chemical definition used by chemists, is any exothermic reaction between a fuel and an oxidizer that produces heat and, typically, an oxidized product. When wood burns, carbon and hydrogen in the wood combine rapidly with oxygen in the air, releasing energy and producing carbon dioxide and water vapor. When iron rusts, iron atoms combine with oxygen, usually with water playing a facilitating role, releasing energy and producing iron oxide. The reactants are different. The fundamental category of chemical event, oxidation releasing energy, is identical.

The balanced chemical equation for the primary rusting reaction is written as four iron atoms reacting with three oxygen molecules to produce two molecules of iron three oxide, commonly known as rust. This reaction has been measured to release approximately 1829 kilojoules of energy for every four moles of iron consumed, according to standard thermochemical reference values used throughout chemistry education and industrial corrosion science. To put that number in perspective, burning a single mole of methane, the primary component of natural gas, releases about 890 kilojoules. Rusting four moles of iron, roughly two hundred and twenty three grams of iron, releases more total energy than burning a mole of natural gas. The iron in a large steel beam, fully rusted, has released, or will eventually release, a genuinely significant quantity of chemical energy, comparable to a meaningful quantity of fuel.

The difference, as established in the previous chapter, is entirely about rate. Rusting proceeds slowly at room temperature because the reaction, while thermodynamically favorable, faces a real activation energy barrier and typically requires the mediating presence of water and dissolved electrolytes to proceed at a meaningful pace. The full mechanism is more involved than the simple equation suggests. Iron first loses electrons to become iron two plus ions in a reaction facilitated by moisture, those ions react further with dissolved oxygen to form iron three compounds, and the final product, hydrated iron oxide, is what we recognize visually as reddish brown rust. Each of these intermediate steps happens at a measured, unhurried pace, governed by how quickly ions can move through a thin film of moisture on a metal surface, how much dissolved oxygen is available, and how much surface area is exposed.

Because the reaction proceeds through this slow, multi step ionic pathway rather than through rapid gas phase combustion, the energy it releases escapes gradually, as low grade heat, dispersing into the surrounding air or water almost as fast as it is produced. This is precisely why nobody has ever needed a fire extinguisher for a rusting fence post. The power output, meaning the rate of energy release, is so low that any sensitive thermometer pressed directly against actively rusting metal will show only a marginal, often undetectable, temperature increase above ambient conditions.

The Experiment That Proves the Point

You do not need specialized laboratory equipment to observe rust behaving as the exothermic reaction it truly is. The American Physical Society has documented a simple, widely reproduced classroom demonstration, commonly used in science education programs, in which fine steel wool, which has a dramatically larger surface area relative to its mass than a solid piece of iron, is combined with a small amount of table salt, which acts as an electrolyte to accelerate the ionic reactions involved in rusting, and a bit of moisture. Packed into an insulated container with a thermometer, this accelerated rusting reaction can raise the internal temperature of the container by a measurable, sometimes striking, number of degrees within an hour, occasionally reaching levels warm enough to be uncomfortable to the touch. This demonstration works precisely because it does two things simultaneously: it dramatically increases the surface area available for the reaction, following the same principle discussed in the previous chapter regarding dust explosions, and it uses salt to speed up the underlying electrochemical process. Neither change alters the fundamental chemistry. Both changes simply move the reaction rate higher along the spectrum described earlier, from a process that would ordinarily take years toward one that takes an hour.

This experiment has practical applications beyond the classroom. It is, in essence, the underlying scientific principle behind commercial air activated hand warmers, the kind hikers and outdoor workers use in cold weather. These products contain iron powder, salt, activated carbon, and a controlled amount of water, sealed inside a permeable pouch. When exposed to air, the iron begins to oxidize rapidly, releasing heat through exactly the same fundamental reaction that rusts an old bridge, just accelerated by orders of magnitude through fine iron powder and electrolytic assistance, and sustained at a comfortable, controlled temperature for several hours. The chemistry connecting a rusting shipwreck and a pocket hand warmer is, quite literally, identical.

Corrosion as Global Economic Force

Because rust is slow and silent, society has historically underinvested in understanding and preventing it, right up until the financial cost became impossible to ignore. Corrosion engineering, as a distinct discipline, exists largely because the cumulative economic toll of this silent reaction is staggering.

The scale of this cost is difficult to overstate. Studies conducted by corrosion engineering organizations, including detailed national assessments carried out for government infrastructure agencies, have repeatedly estimated that corrosion related costs run into the hundreds of billions of dollars annually within a single large national economy, once you account for direct replacement costs, preventive maintenance, protective coatings, and the downstream costs of corrosion related failures in bridges, pipelines, vehicles, ships, and industrial equipment. Bridges collapse, on rare but well documented occasions, because corrosion silently weakened a critical structural component over years or decades without producing any dramatic warning sign until the moment of failure. Pipelines rupture, sometimes catastrophically, because internal corrosion thinned a pipe wall below its safe operating pressure threshold long before any external inspection revealed the problem.

This is the defining hazard pattern of silent explosions generally, and it is worth stating plainly here because it recurs throughout this book. A loud explosion announces itself and then is over. A silent explosion can proceed for years, doing genuine structural damage the entire time, without ever producing a signal that human senses are naturally equipped to detect. The danger is not that rust releases less total energy or causes less total damage than fire. Often it causes more damage, measured over a long enough timeframe. The danger is that the damage accumulates invisibly, and by the time it becomes visible, in the form of a rust colored stain, a structural crack, or in the worst cases, sudden catastrophic failure, the underlying reaction has often been progressing unnoticed for a very long time.

Why Some Metals Rust and Others Do Not

Not every metal undergoes this kind of silent, destructive oxidation, and the reasons why illuminate the underlying chemistry further. Aluminum, for instance, oxidizes even more readily than iron in a purely thermodynamic sense, yet aluminum objects do not crumble into powder the way rusted iron eventually does. The difference lies in the physical structure of the oxide layer each metal forms. When aluminum oxidizes, it forms a thin, dense, tightly adherent layer of aluminum oxide on its surface, and critically, this layer is essentially impermeable to further oxygen and moisture penetration. Once that microscopically thin protective layer forms, typically within moments of a fresh aluminum surface being exposed to air, it seals the underlying metal off from further reaction almost entirely. This is why aluminum window frames, aircraft panels, and beverage cans can last for decades without significant corrosion, despite aluminum being, chemically, more reactive with oxygen than iron.

Iron oxide, unfortunately, does not share this protective quality. The crystal structure of rust is porous and flaky, and critically, it occupies a larger volume than the original iron it replaced. As rust forms on an iron surface, this volume expansion causes the oxide layer to crack and flake away, continuously exposing fresh, unprotected iron underneath to further oxidation. This is why rust, once it begins, tends to be self perpetuating and progressive, eating steadily deeper into a piece of iron or steel over time, while aluminum oxidation is self limiting and essentially stops after the initial protective layer forms. Engineers describe this critical property using a ratio called the Pilling

Bedworth ratio, which compares the volume of the oxide produced to the volume of the original metal consumed, and it is one of the most important predictive tools in corrosion science for determining whether a given metal will form a protective oxide layer or a destructive, self perpetuating one.

Stainless steel exploits a related principle deliberately. By alloying iron with a meaningful percentage of chromium, typically at least ten and a half percent by modern industry standards, manufacturers ensure that the chromium, rather than the iron, preferentially oxidizes first when exposed to air. Chromium oxide, like aluminum oxide, forms a thin, dense, self healing protective layer, and as long as that layer remains intact or can reform quickly when scratched, the underlying iron in the steel is shielded from the atmospheric oxygen and moisture that would otherwise rust it. This is the entire secret behind why stainless steel resists rusting while ordinary carbon steel does not, despite both being made primarily of iron. It is not that stainless steel is chemically immune to oxidation. It is that stainless steel has been deliberately engineered to oxidize in a self limiting, protective way rather than a progressive, destructive one.

The Statue of Liberty's Hidden Corrosion Problem

One of the most instructive real world illustrations of silent corrosion involves the Statue of Liberty, and the problem is rooted precisely in the metallurgical principle just described. The statue's outer skin is copper, a metal that forms its own protective patina over time, the greenish surface layer visible on the statue today, which developed naturally over decades of atmospheric exposure and now serves a genuinely protective function similar to the oxide layers on aluminum and stainless steel.

The internal structural framework, however, was originally built using iron ribs, designed by the engineer Gustave Eiffel, connecting to the copper skin. Over the decades following the statue's 1886 dedication, a slow, silent, and ultimately serious corrosion problem developed at the junctions between the iron framework and the copper skin, driven by a phenomenon called galvanic corrosion. When two dissimilar metals are in electrical contact in the presence of moisture, one metal will corrode preferentially, essentially sacrificing itself to protect the other, a principle that is sometimes used deliberately in engineering through sacrificial anodes, but which becomes destructive when it happens unintentionally in a load bearing structure. In the Statue of Liberty's case, the iron framework corroded steadily at these junction points for roughly a century, invisible from the outside because it was happening beneath the copper skin, until a major restoration project in the 1980s discovered the extent of the damage and replaced the corroded iron framework with corrosion resistant stainless steel, specifically chosen and electrically insulated from the copper skin to prevent the same galvanic reaction from recurring.

This is a nearly perfect illustration of everything this chapter has been building toward. A structural failure process, driven by real, measurable chemical energy release, proceeded for approximately a century, entirely silently, entirely invisibly, inside one of the most visited and most photographed monuments on Earth, without a single alarm bell, without a wisp of smoke, without anything a human observer could detect without deliberately looking for it.

Living With a Slow Fire

Understanding rust as a genuine, if extremely slow, form of combustion changes how you think about maintenance, engineering, and even history. The rusticles hanging from the SS Central America, the corroded iron ribs once hidden inside the Statue of Liberty, the hundreds of billions of dollars spent annually worldwide fighting corrosion, all of it traces back to the same simple chemical reaction described at the start of this chapter: iron, oxygen, and water, quietly releasing energy, one electron transfer at a time, at a pace too slow for any human sense to register directly.

Modern engineering has developed genuinely effective countermeasures, protective coatings, cathodic protection systems that use a sacrificial metal to absorb the corrosion instead of the structure being protected, alloying strategies like stainless steel that engineer a self limiting oxide layer, and increasingly sophisticated inspection technologies that can detect the early stages of corrosion long before it becomes visible to the naked eye. But every one of these countermeasures exists because engineers eventually took seriously something that, for most of human history, seemed too slow, too quiet, and too undramatic to warrant serious concern. Rust does not announce itself with a bang. It does not need to. Given enough time, it can bring down what fire never touched, and it does so in near perfect silence, releasing genuine chemical energy the entire time, following exactly the same fundamental physics described in the previous chapter, just stretched across a timescale that our senses were never built to perceive.

The next chapter moves from the slowest chemical fire on Earth to something slower still, and considerably more energetic per gram: the silent furnace inside a radioactive atom, and the spacecraft, pacemakers, and scientific instruments that have relied on its quiet, steady heat for longer than many of their operators have been alive.

CHAPTER FOUR

The Radioactive Furnace

Voyager 1 is, at the time of this writing, more than fifteen billion miles from Earth, farther from home than any human made object has ever traveled, moving through interstellar space where the sun is little more than an unusually bright star. It launched in 1977. Its instruments have continued transmitting data back to Earth for over four and a half decades, powered the entire time not by solar panels, which would be nearly worthless that far from the sun, and not by any conventional battery, which would have depleted decades ago, but by a small, sealed canister of plutonium 238 that has been quietly, continuously converting radioactive decay into electricity since before many of its current mission engineers were born.

This chapter is about that canister, and about the strange, powerful, genuinely silent category of energy release happening inside it: nuclear decay.

What Radioactive Decay Actually Is

Every atom is defined by the number of protons in its nucleus, but many elements exist in multiple versions, called isotopes, that share the same number of protons while differing in their number of neutrons. Some of these isotopic configurations are stable indefinitely. Others are not. An unstable nucleus, sometimes called a radioactive nucleus, is one where the internal arrangement of protons and neutrons is not in its lowest possible energy state, and given time, it will spontaneously rearrange itself into a more stable configuration, releasing the difference in energy in the process.

This spontaneous rearrangement is what we call radioactive decay, and it comes in several distinct types. Alpha decay, the type relevant to plutonium 238, occurs when an unstable nucleus ejects a particle consisting of two protons and two neutrons bound together, identical in composition to a helium nucleus, called an alpha particle. This ejection reduces the atomic number of the original nucleus by two and its mass number by four, transforming plutonium 238 into uranium 234, which is itself radioactive and will eventually continue decaying further down a long chain toward a stable isotope of lead. Each individual decay event releases a specific, fixed quantity of energy, determined entirely by the difference in nuclear binding energy between the parent and daughter configurations. For plutonium 238's primary alpha decay pathway, that energy release has been measured at approximately 5.593 million electron volts per decay event, an amount of energy that sounds small stated in those units, but which, when multiplied across the trillions upon trillions of individual atoms decaying every second within even a modest sample, adds up to a genuinely substantial and continuous release of thermal power.

Crucially, none of this process involves combustion, oxidation, or any chemical reaction whatsoever. It is happening entirely inside the atomic nucleus, governed by the strong nuclear force and the weak nuclear force rather than by the electromagnetic forces that govern ordinary chemistry. This is why radioactive decay proceeds identically whether the material sits in open air, inside a sealed vacuum, or embedded in solid rock miles underground. It requires no oxygen, no external heat source, no spark. The energy comes from within the atom itself, and nothing external can meaningfully accelerate or slow the rate at which any given radioactive isotope decays, a property that distinguishes nuclear decay sharply from every other silent explosion described in this book, nearly all of which depend heavily on external conditions like temperature, moisture, or surface area.

The Half Life Clock

Every radioactive isotope decays according to a predictable statistical pattern described by its half life, the amount of time required for exactly half of any given sample to decay into its more stable daughter product. Plutonium 238 has a measured half life of 87.7 years. This means that a sample of pure plutonium 238 will have half its original mass converted to uranium 234 after 87.7 years, a quarter of its original mass remaining as plutonium 238 after 175.4 years, an eighth remaining after 263.1 years, and so on, in a smooth exponential decline that, mathematically, never quite reaches zero but becomes negligibly small after several half lives have elapsed.

This exponential decay curve has a direct, practical consequence for anything powered by plutonium 238: the power output declines predictably and gradually over time, following the same exponential curve as the underlying decay. Engineers who designed the RTGs, or radioisotope thermoelectric generators, that power deep space missions have documented that even after twenty years of operation, a plutonium 238 based power source will still be producing approximately eighty five percent of its original power output, according to figures published by researchers and engineers working directly with these systems. This gentle, predictable, entirely passive decline in power is precisely why RTGs remain the preferred power source for missions traveling beyond the reach of useful sunlight, and why spacecraft launched in the 1970s are still, remarkably, transmitting data today.

From Decay Heat to Usable Electricity

The mechanism by which this silent nuclear decay becomes usable electrical power is itself an elegant piece of engineering that involves no moving parts whatsoever, a genuinely rare quality in any power generation system. Inside an RTG, the radioactive material, typically plutonium dioxide in modern American designs, is encased in layers of shielding and structural material and positioned adjacent to an array of thermocouples, devices that exploit a physical principle called the Seebeck effect, in which a temperature difference across certain material junctions generates a small electrical voltage directly, without any mechanical motion at all.

The radioactive core, heated continuously by its own internal decay, maintains a significant temperature differential relative to the cold vacuum of space on the RTG's exterior. In some configurations, the surface temperature of the plutonium 238 fuel element itself can reach as high as 1050 degrees Celsius, according to figures published by engineers who have worked directly with these power sources, while the outer casing, radiating heat into space, remains dramatically cooler. That temperature gradient, maintained continuously and passively by ongoing radioactive decay, drives a steady electrical current through the thermocouple array. The specific numbers involved in NASA's current generation Multi Mission RTG design, used to power the Curiosity and Perseverance Mars rovers, are well documented: approximately 4.8 kilograms of plutonium 238 oxide, distributed across eight individual 290 watt RTG units, producing roughly 2 kilowatts of total thermal power, of which the thermoelectric conversion process yields approximately 110 watts of usable electrical power, alongside about 2.7 kilowatt hours of energy per day.

This entire process, from radioactive nucleus to spacecraft electricity, happens without a single spark, without a single flame, without a single decibel of sound, and it has now sustained continuous operation on some of humanity's most important scientific instruments for periods exceeding forty years. The Voyager missions, launched in 1977, continue to send back data today, and mission engineers have publicly stated that with careful power conservation measures, at least some of the spacecraft's remaining scientific instruments are expected to continue operating into the 2030s, more than half a century after launch, powered the entire time by the same slow, silent nuclear furnace.

A Brief History of a Difficult Element

Plutonium 238 was the very first isotope of plutonium ever discovered, synthesized in December 1940 by the American chemist Glenn Seaborg and his research associates, who bombarded uranium 238 with deuterons, a form of heavy hydrogen nuclei, to produce neptunium 238, which subsequently decayed into plutonium 238. This makes plutonium 238 chemically and historically distinct from the far more famous plutonium 239, the isotope used in nuclear weapons, and it is worth stating clearly, because public confusion on this point has genuinely affected policy discussions over the decades, that plutonium 238 cannot be used to produce a nuclear weapon. Its nuclear properties, specifically the wrong number of nucleons and an unsuitable neutron economy for sustaining a rapid fission chain reaction, make it fundamentally unusable for that purpose, regardless of quantity or purity. Its value lies entirely in its steady, predictable, alpha emitting decay heat, not in any capacity for a rapid chain reaction.

Producing usable quantities of plutonium 238 has historically been a genuinely difficult and expensive undertaking, requiring the irradiation of neptunium 237 with neutrons inside a nuclear reactor, followed by careful chemical separation to isolate the resulting plutonium 238 from other reaction products. The Savannah River Site in South Carolina, operated for decades as part of the United States nuclear weapons production complex before shifting significant capacity toward isotope production for civilian and scientific use, has been documented by the American Chemical Society as having produced the substantial majority of the plutonium 238 used in American RTGs, delivering its first batch in 1960 and ultimately supplying more than three hundred kilograms of the material between 1959 and 1988, out of a total American inventory of roughly three hundred and ten kilograms, with only about ten kilograms sourced from elsewhere. This scarcity, driven by the genuine technical difficulty and cost of production, has periodically constrained NASA's ability to plan deep space missions requiring RTG power, and renewed domestic production efforts in more recent years have specifically aimed at rebuilding a sustainable supply chain for this uniquely useful, uniquely silent power source.

Beyond Spacecraft: RTGs on Earth

While deep space exploration represents the most visible and dramatic application of radioisotope decay heat, it is not the only one. Radioisotope thermoelectric generators, and their simpler cousins, radioisotope heater units, which provide only heat without electrical conversion, have been used for decades in a range of terrestrial and near Earth applications where reliability over long, unattended periods matters more than raw power output. Remote weather stations and navigation beacons in extremely inaccessible locations, including parts of the high Arctic where conventional battery replacement or solar panel maintenance would be logistically difficult or impossible for months at a time, have historically relied on this technology.

Perhaps the most intimate application, historically, involved cardiac pacemakers. Beginning in the late 1960s and continuing through the 1970s, a number of patients received pacemakers powered by small plutonium 238 sources rather than conventional batteries, precisely because the decades long stable power output of plutonium 238 decay meant a single implanted device could theoretically outlast the battery powered alternatives available at the time by a wide margin, sparing patients repeated surgeries to replace a depleted power source. Modern lithium based pacemaker batteries have since improved enough in longevity and safety that nuclear powered pacemakers are no longer manufactured or implanted, but for a period of medical history, patients were quite literally walking around with a small, sealed, continuously operating radioactive furnace embedded in their chest, silently converting nuclear decay into the electrical pulses that kept their hearts beating on schedule, entirely without their conscious awareness of the physics happening centimeters from their own heart tissue.

Why This Furnace Never Runs Out of Fuel, Until It Does

It is worth pausing to appreciate exactly how different this energy source is from every other one examined in this book. Chemical reactions, whether rust, combustion, or battery discharge, involve the rearrangement of electron configurations around atoms, and they proceed at rates heavily influenced by temperature, concentration, surface area, and the presence of catalysts, all factors that can, in principle, be manipulated to speed up, slow down, or entirely halt a reaction. Nuclear decay obeys none of these chemical rules. You cannot slow plutonium 238's decay rate by cooling it down. You cannot speed it up by heating it, applying pressure, or introducing a chemical catalyst. Its half life of 87.7 years is, for all practical purposes, a fixed constant of nature, unaffected by anything short of extraordinary conditions found only in the interior of stars or in specialized particle physics experiments involving processes entirely unrelated to ordinary heating or chemical exposure.

This is precisely why RTGs have proven so extraordinarily reliable for space missions that must operate, entirely unattended, for decades in one of the most hostile environments accessible to human engineering. There is no reaction to accidentally quench, no fuel to run out of in any conventional sense, no combustion process that could be starved of oxygen in the vacuum of space, because the reaction was never dependent on oxygen, or on any external condition at all, to begin with. The energy is, in the most literal sense, baked into the structure of the atom itself, and it will continue emerging, watt by watt, whether anyone is watching, whether the spacecraft is functioning, whether it is orbiting Saturn or drifting silently through interstellar space fifteen billion miles from the nearest sunlit sky.

That is the defining character of this particular silent explosion. It cannot be extinguished. It cannot be sped up. It simply proceeds, atom by atom, exactly on schedule, releasing genuine, substantial, continuously useful energy for as long as the underlying isotope remains, and it does so in a form of silence more absolute than any other reaction examined in this book, because unlike rust, unlike composting, unlike concrete curing, nuclear decay does not even require the presence of a single other substance to proceed. It needs nothing from its environment at all. It simply happens, one nucleus at a time, in perfect isolation, for as long as physics allows.

The next chapter turns from a silent explosion that proceeds for decades in complete stability toward one that, under the wrong conditions, can transform from equally silent stability into genuine catastrophe in under sixty seconds: the lithium ion battery, and the phenomenon battery engineers call thermal runaway.

CHAPTER FIVE
Runaway Cells

At 10:21 in the morning on January 7, 2013, roughly a minute after the last of one hundred and eighty three passengers had disembarked from Japan Airlines Flight 008 at Boston's Logan International Airport, a member of the cleaning crew noticed smoke drifting from the aft cabin of the parked Boeing 787. A mechanic opened the aft electronics bay and found smoke and flames pouring from the auxiliary power unit battery. He tried to extinguish it with a fire extinguisher. It did not go out. Nine days later, an All Nippon Airways 787 experienced a similar battery malfunction in flight and made an emergency landing in Takamatsu, Japan. Within days, the United States Federal Aviation Administration grounded the entire Boeing 787 fleet worldwide, the first time an American aviation regulator had grounded an entire model line since 1979. The National Transportation Safety Board's subsequent investigation concluded that the battery, manufactured by the Japanese company GS Yuasa, had experienced an internal short circuit that triggered what battery engineers call thermal runaway, ejecting flammable material outside the battery's casing and causing the fire.

Nothing about that battery made noise before it failed. No warning bang, no audible click, no sound distinguishable from the ordinary hum of aircraft systems preceded the moment smoke began pouring from the electronics bay. The battery had been sitting there, powering systems normally, for eighteen days and twenty two flight cycles without incident, before an internal defect, invisible from the outside, crossed a threshold and released its stored energy in a rapid, uncontrolled cascade.

This chapter examines exactly how that cascade works, why it is simultaneously one of the most useful and most misunderstood energy release mechanisms in modern technology, and why a device that sits quietly on your nightstand every night contains a genuinely significant quantity of stored chemical energy, released under normal conditions in a slow, gentle trickle, and under abnormal conditions, in a matter of seconds.

What Is Actually Stored Inside a Lithium Battery

A lithium ion battery stores energy through a reversible chemical process involving the movement of lithium ions between two electrodes, typically a graphite based negative electrode and a metal oxide based positive electrode, separated by a thin porous membrane and immersed in a liquid electrolyte, usually a mixture of lithium salts dissolved in organic carbonate solvents. During charging, lithium ions move from the positive electrode through the electrolyte and embed themselves within the layered structure of the graphite negative electrode, in a process called intercalation. During discharge, this process reverses, and the ions flow back, releasing energy that can be captured as usable electrical current along the way.

This design allows lithium ion batteries to pack an unusually large quantity of energy into a very small volume and a very light package, which is precisely why they have become the dominant battery technology for smartphones, laptops, electric vehicles, and increasingly, commercial aviation, where every kilogram of weight carries a direct fuel cost. But this same high energy density is inseparable from the risk this chapter is about. A material capable of storing a large quantity of chemical energy in a small space is, definitionally, a material capable of releasing a large quantity of energy quickly if something interferes with the controlled discharge process the battery was designed around.

The Four Stages of a Silent Catastrophe

Battery safety researchers and engineers who study thermal runaway typically describe the process in four distinct stages, and understanding each one reveals exactly why these failures so often catch people off guard, giving little to no external warning until the process is already well underway.

The first stage is the onset. Under normal operation, a lithium ion battery generates a small, manageable amount of heat through ordinary electrical resistance, described by the well known physics relationship where heat generation scales with the square of the current flowing through the cell. This heat is easily dissipated under normal conditions.. But if something disrupts this equilibrium, an internal short circuit caused by manufacturing debris, physical damage from a puncture or crush, an external heat source, or overcharging that pushes the cell beyond its designed voltage limits, the cell's temperature can begin rising past a critical threshold called the onset temperature. Laboratory research using accelerating rate calorimetry, a specialized instrument designed to detect the earliest stages of self heating in a sample, has documented that the solid electrolyte interface, a thin protective layer that naturally forms on the graphite electrode during normal battery operation, begins to chemically decompose at temperatures in the range of ninety to one hundred and twenty degrees Celsius, according to research published in peer reviewed battery safety literature. This decomposition is itself exothermic, meaning it releases additional heat, which is precisely the seed of the runaway process to come.

The second stage is acceleration. Once the solid electrolyte interface layer begins breaking down, it exposes the highly reactive lithium intercalated graphite underneath to direct chemical contact with the liquid electrolyte, a contact the protective layer had previously prevented. This newly exposed reactive surface begins reacting exothermically with the electrolyte itself, generating heat at a rate that battery researchers have measured, in controlled laboratory conditions, at approximately three hundred and fifty joules per gram of active material when temperatures sit between one hundred and one hundred and twenty degrees Celsius. Crucially, this newly generated heat further accelerates the very same reaction that produced it, in a self reinforcing feedback loop. The rate of self heating during this stage has been documented to increase in an almost linear relationship with rising temperature, and once that self heating rate crosses a documented threshold, roughly zero point two degrees Celsius per minute according to battery safety research, engineers consider the cell to have entered the point of no return, where internal chemistry alone, independent of any external heat source, will continue driving the temperature upward.

The third stage is runaway itself, the phase from which the process draws its name. As temperature continues climbing, the separator membrane, the thin porous barrier physically preventing the positive and negative electrodes from touching each other directly, begins to melt, typically somewhere in the range of one hundred and thirty to one hundred and seventy degrees Celsius depending on the specific separator material used. Once this separator fails, the electrodes can make direct physical contact, causing a massive internal short circuit that dumps the cell's entire remaining stored electrical energy directly into heat in a very short span of time. Simultaneously, the electrolyte itself begins decomposing at these elevated temperatures, releasing flammable gases, including hydrogen, carbon monoxide, and various hydrocarbon vapors, which build pressure inside the sealed cell casing.

The fourth and final stage is destruction. The combination of massive internal short circuiting, rapid electrolyte decomposition, and building internal gas pressure typically culminates in the cell's safety vent rupturing, releasing the accumulated hot, flammable gas suddenly into the surrounding environment, frequently igniting on contact with ambient oxygen and producing the visible flame, hissing sound, and occasionally forceful ejection of material that most people associate with a battery fire or explosion. Research examining cylindrical eighteen six fifty format lithium cells, one of the most common and well studied battery formats, found that at a constant twenty watt external heating rate, cells that failed under higher charging currents released gas violently enough to eject not just gas but liquid electrolyte material as mass, whereas lower current failures tended to release primarily gas without significant liquid ejection, illustrating how the specific conditions leading up to a thermal runaway event directly shape how violent its final, visible stage becomes.

What is critical to understand, and what makes this entire four stage process so genuinely dangerous from a practical safety standpoint, is that stages one and two, which together can last anywhere from several seconds to several minutes depending on the specific failure mode and cell chemistry, produce almost no external warning signs a person could reasonably detect without specialized monitoring equipment. There is no sound. There may be a very slight warmth to the touch, easily mistaken for the ordinary warmth of normal charging. By the time stage three begins, and certainly by stage four, the process has already progressed well beyond any point where a bystander could realistically intervene to stop it.

Why the Note 7 Became a Global Case Study

If the Boeing 787 incidents demonstrated thermal runaway's implications for aviation safety, the Samsung Galaxy Note 7 crisis of 2016 demonstrated its implications for consumer trust and corporate accountability at an almost unprecedented scale. Samsung's own internal investigation, following an extraordinary effort involving seven hundred researchers and engineers who tested more than two hundred thousand devices and more than thirty thousand individual batteries in an attempt to replicate the failures reported by customers, ultimately identified two separate and distinct manufacturing defects, present in batteries sourced from two different suppliers, both of which triggered thermal runaway through slightly different mechanisms.

The first defect involved batteries with edges that had been crimped in a way that allowed internal electrode contact, creating a short circuit that rapidly heated the cell and triggered runaway in surrounding cells within the same battery pack. After Samsung recalled the initial batch of affected phones and issued replacement units using batteries from a different supplier, a second, unrelated defect emerged, this time involving rough edges on the cathode material that similarly caused internal short circuits and thermal runaway. The fact that two entirely independent manufacturing flaws, from two different suppliers, both produced the same fundamental failure mode, offers a striking illustration of just how many different physical pathways can lead a lithium ion cell into the same four stage runaway process described above. A crimped edge and a rough cathode surface are, superficially, completely different manufacturing defects. Both, ultimately, did the same thing: they allowed internal contact between components that were never supposed to touch, providing a low resistance pathway for a sudden, large scale short circuit.

The financial and logistical scale of the Note 7 crisis was considerable. Samsung ultimately issued a total recall of the device, discontinuing it entirely, at an estimated cost, according to industry analysis cited in peer reviewed research on lithium ion battery safety, of approximately seventeen billion dollars once the full scope of recalled units, replacement costs, and reputational damage were accounted for. The recall itself, from a pure logistics standpoint, was remarkably successful, with over ninety six percent of all affected units ultimately returned to Samsung, according to figures reported in industry analysis of the incident. Airlines worldwide, in the wake of widely reported incidents of replacement Note 7 units also catching fire, took the additional step of specifically banning the device from being brought onto flights at all, whether in checked baggage or carried by a passenger, a restriction that remained in place for an extended period and represented one of the first instances of a specific consumer electronics product being individually named and banned in aviation safety regulations.

State of Charge and the Physics of Risk

One of the more counterintuitive findings to emerge from battery safety research involves the relationship between how full a battery is, referred to by engineers as its state of charge, and how dangerous a potential thermal runaway event becomes. Research conducted using accelerating rate calorimetry on cylindrical lithium ion cells found that batteries with a higher state of charge tended to exhibit a lower onset temperature for thermal runaway, meaning fully charged batteries begin the runaway process at a slightly lower temperature than partially discharged ones. At the same time, once runaway does begin, batteries with a higher state of charge, up to a documented threshold of around eighty percent, tend to reach a higher final peak temperature and release energy more violently, because there is simply more stored electrical and chemical energy available to convert into heat during the cascading failure.

This is precisely why aviation safety guidance, shipping regulations for lithium batteries, and increasingly, consumer safety recommendations for storing devices you will not use for an extended period, often suggest keeping batteries at a partial rather than full charge when they will sit unused for a significant length of time. It is also part of the underlying physical logic behind battery management systems in modern electric vehicles and consumer electronics, which are deliberately engineered to avoid charging a cell to its absolute maximum voltage capacity in normal daily use, sacrificing a small amount of total available capacity in exchange for a meaningfully lower baseline risk profile across the millions of charge cycles a battery will undergo over its working life.

Engineering the Silence Back In

The lithium battery industry has spent the years since these high profile incidents developing increasingly sophisticated countermeasures, and studying them reveals just how many different points in the four stage runaway process can, in principle, be interrupted before catastrophic failure occurs. Modern separator materials increasingly incorporate ceramic coatings specifically designed to resist melting at the temperatures where conventional polymer separators would fail, extending the window before electrodes can make direct contact. Battery management systems now typically monitor individual cell voltages and temperatures within a larger battery pack in real time, capable of detecting the earliest, most subtle signs of an individual cell beginning to deviate from normal behavior long before a human observer could notice anything wrong, and isolating that cell electrically before its failure can propagate to its neighbors.

Some newer battery chemistries offer inherently improved safety margins. Lithium iron phosphate cells, increasingly used in electric vehicles and stationary energy storage systems, have been documented in peer reviewed battery safety research to exhibit a dramatically higher thermal decomposition onset temperature, in the range of five hundred to eight hundred degrees Celsius, compared to the roughly one hundred to two hundred degree range typical of the nickel cobalt manganese chemistry involved in many of the consumer electronics incidents described above. This chemistry also releases meaningfully less total heat during decomposition and, according to the same research, tends to maintain its structural integrity even after undergoing irreversible phase transitions at extreme temperatures, rather than collapsing entirely. This represents a genuine engineering tradeoff: lithium iron phosphate cells typically offer somewhat lower energy density than the chemistries involved in the Boeing and Samsung incidents, meaning more weight or volume is needed to store the same quantity of energy, but in exchange, they offer a substantially wider safety margin against thermal runaway, which is precisely why this chemistry has become increasingly dominant in applications like electric vehicles and grid scale energy storage, where the sheer number and size of battery cells involved makes even a small per cell risk of runaway a serious aggregate concern.

Living With a Pocket Sized Furnace

The device charging on your nightstand tonight, whatever it is, contains a genuine, meaningful quantity of stored chemical energy, released under normal circumstances in the gentle, controlled trickle that powers your screen and your phone calls, and under abnormal circumstances, capable of releasing that same total energy in a matter of seconds through the four stage cascade described in this chapter. This is not a reason for alarm about ordinary device use. The overwhelming majority of lithium ion batteries, across billions of devices manufactured and used daily worldwide, complete their entire operational lifespan without ever entering thermal runaway, precisely because of the extensive engineering safeguards, manufacturing quality controls, and battery management systems developed in direct response to incidents like the ones described in this chapter.

But understanding the mechanism matters, because it explains something genuinely important about the nature of silent explosions generally, a theme that will recur throughout the remainder of this book. The most dangerous silent energy releases are rarely the ones that are silent and stable indefinitely, like the radioactive decay examined in the previous chapter. They are the ones that are silent and stable until a specific threshold is crossed, at which point the silence ends abruptly and catastrophically, often with almost no warning between the stable state and the dangerous one. A lithium battery sitting quietly on a table is, chemically speaking, always just a short circuit away from becoming a very different kind of event, and the entire discipline of modern battery engineering exists to keep that short circuit from ever happening, and to contain the consequences swiftly and safely on the rare occasions when it does.

The next chapter turns from engineered energy storage to something considerably older and, in its own way, considerably stranger: the biological process by which an ordinary pile of hay, grass clippings, or agricultural waste can generate enough internal heat, entirely through the metabolic activity of microorganisms, to set itself on fire from the inside, with no external ignition source required at all.

CHAPTER SIX

The Compost Heap That Cooked Itself

Every year, agricultural extension offices across the farming regions of North America and Australia issue the same warning, and every year, barns burn anyway. The warning concerns hay, an unremarkable, sun dried plant material that most people would never think to associate with fire risk in the absence of a match, a spark, or an electrical fault. Yet hay fires occur with predictable regularity, most often within the first two to six weeks after baling, and investigators who examine the wreckage afterward frequently find no external ignition source at all. No frayed wiring. No discarded cigarette. No lightning strike. The fire began, quite literally, from the inside of the hay itself, a phenomenon farmers have called spontaneous combustion for well over a century, even though the word spontaneous slightly overstates the case. There is nothing random or inexplicable about it. It is a well understood, three stage chemical and biological process, and this chapter walks through it in detail, because it offers one of the clearest illustrations in this entire book of a genuinely silent explosion escalating, in slow motion, toward a genuinely loud one.

Stage One: The Living Pile

Freshly cut hay is not dead plant material in any biologically meaningful sense at the moment it is baled. Plant cells continue respiring for a period after cutting, consuming stored sugars and releasing carbon dioxide and a modest amount of heat, exactly as living plant tissue does. This initial respiration phase, on its own, rarely generates enough heat to pose any danger. The real trouble begins when hay is baled with excess moisture, typically defined by agricultural extension research as moisture content above roughly twenty percent, which creates ideal conditions for a much more vigorous heat source: microorganisms.

Bacteria and fungi naturally present on plant material find damp, baled hay to be an excellent growth medium, rich in carbohydrates and protected from wind and sun by the density of the bale itself. Research conducted at Kansas State University, tracking temperature changes in stored hay bales over time, documented that mesophilic bacteria, the first wave of microbial colonizers, thrive particularly well once hay temperature reaches approximately sixty eight degrees Fahrenheit, a temperature easily achieved by ordinary ambient warmth combined with the initial heat of plant respiration. As these bacteria multiply and metabolize the sugars and carbohydrates within the damp hay, they release metabolic heat as a byproduct, exactly the same fundamental principle by which your own body generates heat through metabolism, a topic explored in greater depth in a later chapter of this book.

This microbial heat generation, on its own, is enough to begin raising the internal temperature of a hay bale into a genuinely concerning range within a matter of days. Extension research from Virginia and Oklahoma agricultural programs documents that hay stored with sufficient moisture to maintain high internal humidity commonly reaches internal temperatures of one hundred and fifty eight degrees Fahrenheit within a timeframe ranging from a few days, under warm and humid conditions, to several weeks, under cooler and drier conditions. At this temperature, something important happens: thermal death begins killing off the mesophilic bacteria responsible for the initial heating. You might expect this to be the end of the danger, since the organisms generating the heat are dying. It is, instead, precisely where the process becomes genuinely hazardous.

Stage Two: When Biology Hands Off to Chemistry

Once bacterial activity begins dying back due to the heat it generated, hay does not simply cool down. Instead, research summarized by Oklahoma State University's agricultural extension program describes a critical transition: once internal temperature climbs into the range of roughly one hundred and thirty to one hundred and seventy five degrees Fahrenheit, a purely chemical, non biological exothermic reaction can begin, entirely independent of any living organism. This reaction involves oxidative chemistry occurring directly between atmospheric oxygen, penetrating into the bale through natural air gaps, and various organic compounds within the plant material itself, and unlike the microbial heating that preceded it, this chemical oxidation does not require living organisms to sustain itself. It can, and does, continue accelerating even as the last of the bacteria that started the process die off from the very heat their metabolism produced.

This handoff from biological to chemical heating is the single most important, and most dangerous, feature of the entire spontaneous combustion process, because it means the danger does not end when the microorganisms stop being active. If anything, this second stage tends to accelerate faster than the first, precisely because chemical oxidation reactions generally follow the same exponential relationship between temperature and reaction rate described in chapter two of this book. Higher temperature drives faster oxidation, which releases more heat, which drives temperature higher still, in the same self reinforcing feedback loop that characterizes thermal runaway in lithium batteries, examined in the previous chapter, and the same basic feedback structure underlying nearly every dangerous silent explosion covered in this book.

Farm safety researchers have documented that hay treated with certain chemical preservatives, specifically compounds containing ethoxyquin and butylated hydroxytoluene, can begin releasing hydrogen cyanide gas once internal temperatures reach approximately one hundred and fifteen degrees Celsius, adding a serious toxic hazard on top of the fire risk itself for firefighters and farm workers who might otherwise assume they are dealing with a straightforward, if dangerous, hay fire.

Stage Three: Ignition

The final stage of this process is the one that produces the visible, audible fire most people associate with the phrase spontaneous combustion. As the chemical oxidative reaction described above continues accelerating, hay's internal temperature can climb into a range where actual open combustion becomes possible, typically cited by multiple agricultural extension sources as falling between four hundred and forty eight and five hundred and twenty seven degrees Fahrenheit. Once ignition occurs, the fire behaves largely as any ordinary hay or straw fire would, spreading through the dry outer material of the bale or stack with the speed and violence typical of any large fuel load igniting in the open presence of oxygen.

What is worth pausing on is just how long a journey this represents, from a damp, freshly cut bale of hay sitting quietly in a barn, to an inferno capable of destroying an entire agricultural structure. Agricultural safety researchers place the typical total timeline for this progression, from initial baling to eventual ignition, at somewhere between four and ten weeks, according to Virginia Cooperative Extension research, though the more aggressive cases documented by other extension programs describe ignition occurring within the first two to six weeks after baling. Farmers who have lost barns to this process frequently report that nothing appeared visibly wrong until flames were already visible, precisely because the entire multi week buildup, encompassing both the microbial heating stage and the subsequent chemical oxidation stage, happens entirely inside the dense interior of a hay bale or stack, insulated from view and, for most of the process, from smell or any other external sensory signal.

A tragic and well documented illustration of the real world stakes involved comes from agricultural safety reporting on a barn fire roughly thirty minutes outside a small community in New York State, in which spontaneous hay combustion destroyed the structure and killed eighteen cows housed inside. No external ignition source was found. The cause, as determined by investigators and reported in agricultural safety literature examining the case, was the same three stage process described throughout this chapter: damp hay, insufficiently dried before baling, generating first biological and then chemical heat, entirely undetected, until the accumulated energy crossed the threshold into open flame.

Why Size and Density Matter So Much

Two physical factors determine, more than almost anything else, whether a given batch of damp hay will progress harmlessly through some mild warming and then cool back down, or continue all the way to ignition: the size of the bale or stack, and how densely it is packed.

Hay is, among its other properties, an excellent thermal insulator, which is precisely why it has been used historically to insulate everything from farm buildings to, in a curious historical footnote, early refrigeration efforts before mechanical cooling existed. This insulating property works against safety in the context of a large hay mass generating internal heat, because it means that heat produced deep within a large bale or stack has genuine difficulty escaping to the surrounding air, where it could dissipate harmlessly. The larger the mass of hay involved, the smaller its surface area is relative to its total volume, and the harder it becomes for internally generated heat to reach the surface and radiate away before it accumulates to a dangerous level. This is precisely the same surface area to volume principle discussed in chapter two regarding why finely divided materials react faster than solid blocks of the same substance, simply operating in reverse: a small, loose pile of hay has abundant surface area relative to its volume, allowing heat to escape easily, while a large, tightly packed round bale or an entire barn stack has comparatively little surface area relative to its volume, trapping heat efficiently deep in its core.

This is precisely why agricultural safety guidance consistently recommends storing suspect hay, meaning hay baled at higher than ideal moisture content, in loose stacks or rows outdoors rather than tightly packed inside an enclosed barn, specifically to maximize air circulation and surface area for heat dissipation, and to ensure that if ignition does eventually occur despite these precautions, the resulting fire causes far less total damage than the same event happening inside a fully enclosed structure packed with additional fuel in the form of surrounding hay, wooden beams, and often valuable equipment or livestock.

The Monitoring Solution

Because this entire multi week process is invisible from the outside for most of its duration, and because by the time visible smoke or smell becomes apparent the situation has often already progressed dangerously close to ignition, modern agricultural safety practice increasingly relies on direct temperature monitoring rather than passive observation. Long probe thermometers, designed specifically to be driven deep into the core of a hay bale or stack where the danger actually originates, allow farmers to track internal temperature over the critical first several weeks after baling, when the risk is highest. Extension guidance across multiple agricultural programs establishes a consistent set of action thresholds: temperatures approaching one hundred and forty degrees Fahrenheit warrant close, frequent monitoring, temperatures reaching one hundred and fifty degrees Fahrenheit are considered a serious warning sign that the exothermic chemical oxidation stage may already be underway, and at that point, agricultural safety guidance consistently and explicitly warns that internal temperature will typically not decrease on its own, meaning spontaneous combustion should be considered essentially imminent rather than merely possible.

More recently, agricultural technology companies have introduced wireless monitoring systems that embed temperature sensing probes directly into stored hay and transmit real time readings to a farmer's phone or computer, allowing continuous, twenty four hour monitoring without requiring someone to physically walk out to a barn and manually check a thermometer probe at regular intervals throughout the day and night, a genuine improvement given that dangerous heating can develop and progress significantly within a period as short as three to seven days under the wrong conditions of moisture and ambient warmth.

The Broader Pattern This Chapter Reveals

What makes hay fire such an instructive case study, within the broader argument of this book, is how cleanly it demonstrates the full arc from silent to loud, moving through every stage described in chapter two: an initial reaction well below the activation energy needed for open combustion, proceeding slowly and invisibly through biological and then chemical means, generating genuine measurable heat the entire time without producing sound, flame, or any other obvious external signature, and then, upon crossing a specific measurable temperature threshold, transitioning within a comparatively short final window into the loud, fast, visually dramatic combustion event that everyone would immediately recognize as an explosion or fire in the conventional sense.

Every single stage of that arc, from the first bacterium multiplying in a slightly too damp bale of hay to the final moment of open flame, obeys the exact same thermodynamic and kinetic principles established earlier in this book. There is no special, separate category of physics governing spontaneous combustion. There is simply organic material, moisture, oxygen, and time, combining according to ordinary, well understood chemistry, and the entire multi week process would be entirely predictable, and in most cases entirely preventable, if more people understood that the danger begins long before any flame becomes visible, hidden silently inside a pile of what looks, from the outside, like nothing more than ordinary, harmless, sun dried grass.

The next chapter shifts from agriculture to civil engineering, and examines a silent explosion of truly monumental scale: the heat released when enormous quantities of cement and water combine to form concrete, a reaction responsible for one of the more remarkable and lesser known engineering challenges in the construction of one of the largest concrete structures ever built.

CHAPTER SEVEN

Concrete Fury

Six hundred miles of steel pipe, buried inside a single structure. That is not a typographical error, and it is not a metaphor. When engineers of the United States Bureau of Reclamation began pouring the concrete that would become Hoover Dam in June 1933, they faced a problem that had never been solved at this scale before, and the solution required threading enough steel pipe through the dam's interior to stretch, laid end to end, from roughly Los Angeles to well past Salt Lake City. The problem they were solving had nothing to do with structural reinforcement in the conventional sense. It had to do with heat, an enormous, genuinely dangerous quantity of heat, generated entirely by a chemical reaction that most people walk past every single day without a second thought: the hardening of concrete.

The Chemistry Hiding in Plain Sight

Concrete's transformation from a wet, pourable slurry into a rock hard structural material is not, contrary to what many people assume, simply a matter of the water evaporating and the mixture drying out. It is a genuine chemical reaction, called hydration, in which water molecules chemically combine with the mineral compounds present in Portland cement, the binding ingredient that holds concrete's sand and gravel aggregate together. This reaction is irreversible, and critically for the subject of this book, it is exothermic, releasing real, measurable heat as it proceeds.

Portland cement, the material engineers describe as an overview by materials scientists confirms, is composed primarily of four key mineral compounds, commonly abbreviated using cement industry shorthand as C3S, C2S, C3A, and C4AF, technical names referring to specific combinations of calcium, silicon, aluminum, and iron oxides fused together at extremely high temperature during the cement manufacturing process. When water is introduced, each of these compounds undergoes its own distinct hydration reaction, at its own distinct rate, releasing its own distinct quantity of heat. Tricalcium aluminate, reacting in the presence of gypsum and water to form a compound called ettringite, releases heat at a rate of approximately two hundred and seven calories per gram according to civil engineering research on cement chemistry, and this reaction proceeds rapidly, generating enough heat quickly enough to help trigger the subsequent, slower reactions involving the calcium silicate compounds. Tricalcium silicate, often called alite, hydrates to form calcium silicate hydrate gel, the primary compound responsible for concrete's characteristic strength, releasing approximately one hundred and twenty calories of heat per gram in the process, while the more gradual hydration of dicalcium silicate, or belite, releases a comparatively modest sixty two calories per gram but continues contributing to the concrete's strength gain over a much longer timeframe, sometimes continuing for years after the initial pour.

In an ordinary residential concrete slab, poured a few inches thick, this hydration heat dissipates into the surrounding air and ground almost as quickly as it is generated, exactly the same way the heat from a rusting nail dissipates too quickly to ever become noticeable, as discussed in chapter three. The reaction proceeds, the concrete cures, and nobody standing nearby ever perceives anything unusual happening, because the total power output, the rate of heat release relative to the surface area available for that heat to escape, remains low enough to stay well within what ordinary conduction and convection can carry away harmlessly.

Scale changes everything.

When Concrete Becomes a Furnace

Hoover Dam required, according to figures compiled by construction historians and engineering researchers, more than three point two five million cubic yards of concrete, a volume so large that popular engineering accounts commonly note it would be sufficient to build a standard two lane highway stretching the full distance from San Francisco to New York City. Every single cubic yard of that concrete, as it cured, released hydration heat through precisely the chemical reactions just described. In a structure of ordinary size, this heat would escape harmlessly. In a structure whose core sat many dozens of feet from the nearest exposed surface, the heat had nowhere to go.

Engineers working on the project calculated that if the dam had simply been poured as a single continuous mass and allowed to cure naturally, without any special intervention, the concrete deep in the dam's interior core would have taken approximately one hundred and twenty five years to fully cool to ambient temperature, according to figures cited by multiple engineering historians examining the project. This was not merely an inconvenience that would have slowed the construction schedule, though it certainly would have. It represented a genuine structural danger. As the exterior portions of a massive concrete pour cool and contract at a normal rate while the interior remains hot and expanded, the resulting internal stress, driven by that temperature differential, can crack the concrete catastrophically, precisely the kind of structural flaw that would be entirely unacceptable in a dam designed to hold back the full force of the Colorado River indefinitely.

The engineering solution developed for Hoover Dam became a landmark achievement in construction history in its own right, and it is worth describing in some detail, because it represents one of the clearest examples in this entire book of engineers explicitly, deliberately treating a silent chemical reaction as the genuine, significant hazard it actually was, decades before the word thermal management became commonplace in construction vocabulary.

Rather than pouring the dam as one continuous mass, engineers divided the structure into interlocking blocks, each roughly five feet by five feet by five feet according to the original design, later adjusted during construction. Before each block was poured, workers embedded a network of one inch steel pipes throughout the wet concrete. Once a given block was poured, ice cold water was circulated continuously through this embedded pipe network, physically carrying accumulated hydration heat out of the concrete's interior far faster than passive conduction to the surrounding air ever could have managed. This artificial cooling system relied on a dedicated refrigeration plant built specifically for the project, which according to construction historians produced roughly one thousand pounds of ice daily to support the cooling operation, alongside chilled water circulation. Engineers also pre cooled the raw ingredients, the sand, gravel, and water, before mixing them into concrete in the first place, reducing the total heat load the cooling pipes would eventually need to remove.

The results were dramatic. Rather than the naturally projected one hundred and twenty five years required for the dam's core to cool on its own, each individual five foot block, actively cooled through its embedded pipe network, could be brought down to a safe, stable temperature within a matter of days, according to figures reported by construction industry historians examining the project. Once a given block had cooled sufficiently, workers filled its internal cooling pipes with grout, both sealing them permanently and adding a small amount of additional structural material, before proceeding to pour and cool the next adjacent block. This block by block, actively cooled construction method allowed a structure that would otherwise have taken over a century to safely cure to be built, cured, and made structurally sound within a period of a few years, a genuinely remarkable feat of engineering built entirely around managing an invisible, silent, but very real and very large exothermic chemical reaction.

Why Modern Mega Structures Still Fight This Battle

Hoover Dam is a particularly dramatic historical example, but the underlying challenge it illustrates, generally referred to in modern civil engineering as mass concrete thermal management, remains an active, ongoing concern in large scale construction projects today. Contemporary research published in engineering journals continues to model and refine predictions of hydration heat generation in large concrete pours, using increasingly sophisticated simulations that account for the specific mineral composition of the cement being used, the size and geometry of the structure, ambient environmental conditions, and any thermal control measures being employed. Engineers building large modern dams, particularly those using a technique called roller compacted concrete, still rely on cooling strategies directly descended from the principles pioneered at Hoover Dam, including embedded cooling pipes and carefully sequenced pour schedules designed to prevent any single mass of curing concrete from accumulating dangerous internal heat.

The stakes of getting this wrong remain genuinely serious even in modern construction. Uncontrolled thermal cracking in large concrete structures, driven by exactly the differential cooling stress described above, can compromise structural integrity, allow water infiltration that accelerates long term deterioration through mechanisms like freeze thaw damage and reinforcement corrosion, discussed in chapter three, and in the most severe cases, require extensive and expensive remediation or even partial reconstruction of affected structural elements. Modern engineering research on this topic frequently references what are called temperature rise inhibitors, chemical additives specifically formulated to slow the initial rate of hydration heat release, buying more time for that heat to dissipate naturally or through active cooling systems before it accumulates to dangerous levels, essentially achieving through chemistry a similar outcome to what Hoover Dam's engineers achieved through brute force refrigeration and physical pipe networks.

From Ice Pipes to Algorithms: How Engineers Model Concrete Heat Today

The fundamental physical problem Hoover Dam's engineers faced in 1933, a large mass of curing concrete generating more hydration heat than its surface area could passively dissipate, has never gone away. It has simply become far better understood, and far more precisely managed, through tools those original engineers could scarcely have imagined. Contemporary civil engineering research defines mass concrete formally, following American Concrete Institute guidance, as any concrete structural member with a minimum dimension greater than roughly four feet, a threshold at which the ratio between a structure's volume and its exposed surface area becomes large enough that internally generated hydration heat can no longer escape quickly enough to prevent a dangerous internal temperature buildup, precisely the geometric principle that made Hoover Dam's core so much more thermally hazardous than an ordinary concrete slab.

Where Hoover Dam's engineers relied on physical experimentation, careful record keeping, and hard won empirical judgment to determine how much cooling their embedded pipe network would need to provide, modern engineers increasingly rely on detailed computational simulation before a single cubic yard of concrete is poured. Research published in construction materials journals describes finite element models capable of predicting, with considerable precision, the temperature field that will develop throughout a proposed concrete structure over the days and weeks following its pour, incorporating detailed calorimetry measurements of exactly how much heat a specific cement formulation will release, at exactly what rate, combined with the specific geometry of the structure, the ambient temperature conditions expected during construction, and the placement of any planned cooling infrastructure. One documented case study examined the foundation raft for a VVER 1200 nuclear reactor building, a single monolithic concrete pour totaling 5,240 cubic meters, where engineers used this kind of simulation based approach to design an unconventional pouring sequence, filling the massive foundation from its center outward toward its edges in carefully timed stages, specifically to moderate the temperature gradient that would otherwise develop between the pour's hot core and its cooler perimeter. Field monitoring following construction confirmed the approach worked, keeping the difference between the concrete's core and surface temperature to just under twenty degrees Celsius, comfortably below the twenty five degree threshold research in this field commonly identifies as the point past which thermal cracking risk becomes significant, and post construction inspection found no thermal cracks whatsoever in the completed foundation.

This same computational approach has proven valuable across an increasingly wide range of modern construction projects wrestling with the identical underlying physics Hoover Dam's builders confronted nearly a century ago. Engineers monitoring a mass concrete bridge foundation in Iowa used specialized software, informed by laboratory calorimetry testing of the specific concrete mix involved, to predict the structure's thermal performance before construction began, allowing adjustments to the mix design and pour schedule to be made on paper rather than discovered as a costly problem in the field. Similar monitoring and modeling techniques have been applied to raft foundations beneath high rise buildings in Hong Kong and Saint Petersburg, to offshore wind turbine foundations, and to a wide range of other mass concrete applications where the same fundamental hazard, hydration heat trapped in a large concrete volume with too little surface area to dissipate it passively, continues to threaten structural integrity exactly as it did at Hoover Dam, simply at a scale and in a geometry specific to each individual modern project.

Changing the Chemistry Rather Than Just Managing the Symptom

Alongside these advances in predictive modeling, modern concrete engineering has developed an entirely complementary strategy for managing hydration heat, one that Hoover Dam's engineers, working with the cement chemistry available in the 1930s, did not have available to them: modifying the cement itself to release less heat in the first place, rather than relying solely on external cooling to remove heat after it has already been generated. Research comparing different cementitious materials has documented that ordinary Portland cement typically generates somewhere between two hundred and ninety and five hundred joules of hydration heat per gram, while blended cements incorporating substantial proportions of industrial byproduct materials, particularly fly ash, a fine powder byproduct of coal fired power generation, and ground granulated blast furnace slag, a byproduct of iron and steel manufacturing, can reduce that total heat generation meaningfully, with laboratory studies documenting reductions in early stage hydration heat release in the range of twenty to thirty three percent when these supplementary materials replace a significant portion of the cement content in a given concrete mix.

This represents a genuinely elegant engineering solution, addressing the underlying chemistry described earlier in this chapter directly, rather than simply building more elaborate infrastructure to manage the heat that chemistry produces. Tricalcium aluminate and tricalcium silicate, the compounds responsible for the most rapid and most intense hydration heat release described earlier in this chapter, are present in smaller proportions in these blended cements, since a portion of the mix's total cementitious material has been replaced by fly ash or slag, materials that still contribute meaningfully to the concrete's final strength through their own slower, less exothermic reaction pathways, but that generate substantially less heat, and generate it more gradually, during the critical early curing period when thermal cracking risk is highest. Combined with computational thermal modeling, layered pour sequencing that allows partially cured concrete to begin dissipating its own heat before the next layer is added on top, and, when necessary, the same fundamental embedded cooling pipe strategy Hoover Dam pioneered, still very much in active use in modern mass concrete construction according to current engineering research, today's engineers have a genuinely comprehensive toolkit for preventing the kind of catastrophic thermal cracking that concerned Hoover Dam's builders so seriously nearly a century ago.

What makes this modern engineering story worth including alongside the historical Hoover Dam account is exactly how directly it illustrates a theme running throughout this entire book: a silent hazard, once genuinely understood at the level of its underlying chemistry and physics, becomes systematically manageable, not through fear or avoidance, but through increasingly precise measurement, modeling, and engineering intervention, refined generation after generation as understanding deepens. The specific tools have changed enormously since 1933, moving from ice production plants and empirically sized steel pipe networks to calorimetry laboratories and finite element simulation software. The underlying physical problem, and the underlying engineering philosophy required to solve it, have remained essentially unchanged, a continuous thread connecting the builders of one of the twentieth century's most celebrated engineering achievements to the engineers currently designing the foundations of nuclear power plants and high rise towers being built at this very moment.

An Underappreciated Kind of Power

There is something genuinely striking about the fact that concrete, perhaps the most mundane, unglamorous building material in the modern world, the stuff of sidewalks and parking garages and highway overpasses, is capable of generating enough internal heat, through pure chemistry, to threaten the structural integrity of one of the largest engineered structures ever built by humans, entirely without flame, without any external heat source, and without producing a single sound audible to anyone standing nearby. The chemical reaction responsible, the hydration of Portland cement, proceeds today in every concrete pour happening anywhere in the world, in driveways and foundations and skyscraper cores, releasing genuine, measurable exothermic heat according to the exact same chemistry described at the start of this chapter. In small pours, that heat vanishes into the environment so quickly that it never crosses the threshold of human perception. In pours large enough, it becomes a genuine engineering hazard requiring active, sophisticated intervention to manage safely.

This is, once again, precisely the pattern established in chapter two of this book: identical underlying chemistry, producing dramatically different consequences purely as a function of scale and the resulting ratio between heat generation and heat dissipation. A garden path poured on a Saturday afternoon and the core of Hoover Dam are, chemically, undergoing the exact same reaction. One never generates enough concentrated heat to matter. The other required six hundred miles of steel pipe and a dedicated ice production plant to keep from cracking apart under its own chemically generated internal fever.

The next chapter turns to a category of energy release that operates according to different rules entirely: not the release of energy through chemical bond rearrangement, as in every reaction examined so far, but the release and absorption of energy through physical changes of state, the transitions between solid, liquid, and gas that quietly govern phenomena ranging from the stability of a glacier to the surprising violence hidden inside a can of compressed gas.

CHAPTER EIGHT

The Slow Detonation

At 9:18 in the morning on June 8, 1998, the DeBruce Grain Elevator in Wichita, Kansas, one of the largest grain storage facilities in the world at the time, erupted in a series of explosions that killed seven workers and trapped ten more inside the wreckage. Investigators later determined that the elevator's dust collection system, designed specifically to prevent the exact hazard that destroyed the facility, had completely failed on the day of the disaster. Every dust collection tank had filled to capacity and backed up, rendering the entire pneumatic dust removal system useless, and allowing fine grain dust to accumulate steadily, invisibly, throughout the structure for an unknown period before the explosion occurred. The facility's vertical design, investigators noted, created what engineers call a chimney effect, in which hot gases and suspended dust could rise rapidly through the structure once ignition occurred, dramatically amplifying the destructive spread of what began as a comparatively localized event.

This was not an isolated tragedy. Just over twenty years earlier, in a single catastrophic month spanning December 1977 and January 1978, five separate grain elevator explosions across the United States killed sixty two people and injured fifty three more, according to figures compiled by the United

States Government Accountability Office. The single deadliest of these, at the Continental Grain Company elevator in Westwego, Louisiana, on December 22, 1977, killed thirty six people in an explosion powerful enough, according to contemporary reporting, to be felt for miles around the site. Purdue University's ongoing national survey of grain dust explosions, one of the most comprehensive tracking efforts of its kind, documented seven such explosions across the United States in 2025 alone, resulting in ten injuries and four fatalities, figures that, while lower than the catastrophic years of the late 1970s, demonstrate that this hazard remains genuinely active in modern agricultural and industrial operations even after nearly half a century of dedicated safety regulation and engineering countermeasures.

This chapter uses grain dust explosions as a deliberate contrast case within this book, because they illustrate, with unusual clarity, the exact mechanism by which a genuinely silent, invisible hazard, fine dust accumulating quietly in the corners and rafters of an industrial structure, can transition with almost no warning into one of the loudest, most destructive events described anywhere in these pages. Understanding this transition sharpens everything the rest of this book has established about the relationship between silence and stored energy.

Why Ordinary Grain Does Not Explode

A single kernel of wheat, corn, or any other grain, sitting on a table, poses no explosion hazard whatsoever. You could apply a direct flame to it and it would simply char and burn slowly, exactly the way a log burns slowly, releasing its stored chemical energy at a modest, controlled rate limited by the comparatively small surface area exposed to oxygen relative to the kernel's total mass. This is the same fundamental surface area principle introduced in chapter two, and it is worth restating here because grain dust explosions represent perhaps the single clearest large scale industrial demonstration of that principle in action.

Grain handling operations, by their nature, generate enormous quantities of fine dust as kernels are moved, dropped, augured, and conveyed through elevators, silos, and processing equipment. This dust consists of the same organic material as the grain itself, the same starches and proteins capable of releasing energy through combustion, but ground and abraded down into particles small enough to remain suspended in air for meaningful periods of time. When that same total mass of organic material is divided into particles this fine, the total surface area exposed to atmospheric oxygen increases by an almost unimaginable factor. A single kernel, ground into thousands of individual dust particles, might present a combined surface area many times larger than the kernel's original exterior, each particle now individually and simultaneously exposed to oxygen on every side, rather than only at its exterior surface as before.

This transformation, from a solid kernel that burns slowly and safely to a cloud of fine dust particles suspended in air, each individually surrounded by oxygen, converts a comparatively inert material into something capable, under the right conditions, of combusting almost instantaneously across an entire suspended dust cloud simultaneously, rather than burning progressively from one point outward the way a solid material does. This is precisely the mechanism responsible for every grain elevator explosion described at the start of this chapter, and it is the same underlying mechanism responsible for a range of other industrial dust explosion hazards involving materials as varied as powdered sugar, coal dust, aluminum dust, and various pharmaceutical and chemical powders, anywhere sufficiently fine combustible dust can become suspended in a confined space alongside an ignition source.

The Anatomy of an Escalating Explosion

What makes grain elevator explosions particularly instructive, rather than simply tragic, is the well documented mechanism by which a single, comparatively contained initial explosion can cascade into something dramatically larger and more destructive, a phenomenon investigators and researchers have studied closely in the aftermath of incidents like the ones described above.

The sequence typically begins with what is called a primary explosion, a comparatively localized ignition of an existing dust cloud, perhaps triggered by a mechanical spark from a piece of equipment, friction from a malfunctioning bearing, or in rarer cases, a welding operation conducted without adequate precautions in a dusty environment, a hazard specifically flagged in fire safety literature examining these incidents. This primary explosion, on its own, might cause meaningful but contained damage.

The genuine catastrophe, in nearly every major grain elevator disaster on record, comes from what happens immediately afterward. The pressure wave and turbulence generated by the primary explosion physically disturbs and dislodges accumulated dust that had been resting, quietly, undisturbed, in corners, on ledges, along rafters and equipment surfaces throughout the facility, dust that had built up gradually over days, weeks, or months of ordinary grain handling operations without anyone necessarily recognizing the scale of accumulation. This dislodged dust becomes suddenly, freshly suspended in the air, precisely in the volume immediately ahead of the primary explosion's advancing flame front, creating a dense, freshly available fuel source exactly where the fire is about to travel next. This secondary dust cloud ignites in turn, producing a secondary explosion that is frequently far more powerful and far more destructive than the primary event that triggered it, and this secondary explosion can, in sufficiently dusty facilities, disturb and ignite still further accumulated dust deeper within the structure, in a cascading chain that investigators examining incidents like the DeBruce disaster have identified as the primary reason grain elevator explosions so often escalate from a contained initial event into a structure wide catastrophe.

This is the deflagration to detonation transition mechanism introduced conceptually in chapter two, now grounded in specific, documented, real world incidents. The initial ignition may begin as a relatively modest deflagration, a subsonic flame front propagating through a limited quantity of suspended dust. But the feeding mechanism just described, in which the explosion itself generates the fuel supply for its own escalation by kicking up previously settled dust, can drive the reaction toward detonation, a supersonic pressure wave capable of far greater structural destruction, entirely because the facility had accumulated dust silently, invisibly, in locations nobody was actively monitoring, over a period of ordinary operation stretching back well before the day of the actual incident.

The Silent Buildup Behind Every Loud Disaster

It is worth pausing directly on this point, because it connects grain dust explosions back to the central argument running throughout this book. Every single grain elevator explosion described in this chapter had a genuinely silent phase preceding it, often lasting weeks or months, during which fine combustible dust accumulated steadily on surfaces throughout the facility, entirely without producing any sound, any visible drama, or any signal that would naturally draw a worker's attention to the accumulating hazard. Housekeeping failures, meaning the routine, unglamorous task of regularly cleaning dust accumulation from elevator legs, conveyor housings, rafters, and equipment surfaces, are repeatedly cited across decades of investigation reports as a primary contributing factor in these disasters, precisely because dust accumulation is not, in itself, dramatic or attention grabbing in the way an active fire or a mechanical malfunction would be. A thin, steadily thickening layer of dust on an overhead beam looks, to an untrained eye walking past it during a routine shift, like nothing more than ordinary industrial grime, not like a silently accumulating fuel reserve waiting for a single spark and a single pressure wave to convert it into a secondary explosion capable of collapsing a structure.

This is precisely the same fundamental pattern examined in chapter six regarding spontaneous hay combustion: a hazard that builds silently and gradually, through a process largely invisible to ordinary observation, before eventually crossing a threshold and manifesting as a sudden, dramatic, loud event. The difference in the case of grain dust is that the final trigger, unlike the purely internal chemical threshold crossed inside a hay bale, typically requires a specific external ignition source, a spark, friction, or an open flame, meeting the accumulated dust at exactly the wrong moment. Remove either element, the accumulated dust or the ignition source, and the disaster does not occur, which is precisely why modern grain handling safety regulation focuses so heavily on both fronts simultaneously: rigorous housekeeping standards designed to prevent dangerous dust accumulation in the first place, and strict controls on potential ignition sources, including welding permits, spark resistant equipment design, and careful management of mechanical friction points throughout a facility.

The Regulatory Response and Its Real Impact

The catastrophic losses of December 1977 and January 1978, sixty two deaths across just five weeks, prompted the federal government to commission the National Academy of Sciences to convene a dedicated panel examining the causes and prevention of grain elevator explosions, according to documentation from the Occupational Safety and Health Administration's own historical review of the issue. This panel, completing its work in 1983, produced findings that directly informed the eventual development of OSHA's Grain Handling Standard, a comprehensive federal regulation establishing specific, mandatory requirements around housekeeping, equipment design, worker training, and hazard communication specifically targeted at preventing the kind of cascading dust explosion mechanism described throughout this chapter.

Longer term data compiled by researchers tracking dust explosion incidents across North America suggests these regulatory and engineering interventions have produced a genuine, measurable improvement over time, even though the hazard has never been eliminated entirely. Historical figures covering 1958 through 1976 documented an average of eight grain elevator explosions annually, with six fatalities per year on average, according to a United States Department of Agriculture task force report examined by dust safety researchers. The catastrophic spike of 1977 and 1978, discussed above, prompted the regulatory response that followed. More recent data compiled by the United States Chemical Safety Board, covering general industry incidents from 2006 through 2017, documented an average of nine explosions and five fatalities annually, a meaningfully lower rate than the crisis years of the late 1970s, though, as the Purdue University survey figures for 2025 demonstrate, still very much an ongoing, active hazard rather than a solved historical problem.

What This Chapter Reveals About the Rest of the Book

Grain dust explosions occupy an unusual position within this book, because unlike nearly every other chapter, this one describes events that are undeniably, unmistakably loud, violent, and immediately recognizable as explosions in the most conventional sense of the word. They belong here precisely because the mechanism producing that loud, violent final event is inseparable from a genuinely silent process, the invisible accumulation of fine combustible dust over an extended period, that shares every essential characteristic with the other silent hazards examined throughout this book. The dust does not announce its presence. It does not smell dangerous, look dramatically different from ordinary industrial dirt, or produce any signal that would prompt an untrained observer to treat it as a serious hazard, right up until the moment a spark meets it and the entire silent accumulation converts, in a fraction of a second, into precisely the kind of loud, destructive event that opened this chapter.

This is, in a sense, the darkest lesson in this entire book, though it is also, properly understood, an empowering one. Silent hazards are not inherently less dangerous than loud ones. In the specific case of grain dust, the silence is not a separate category of event from the eventual explosion, it is the explosion's preparatory phase, building the very fuel supply that the final, loud event will consume. Recognizing this connection, understanding that quiet accumulation and dramatic release are two phases of a single continuous physical process rather than two unrelated phenomena, is precisely the shift in perspective that has driven the development of modern grain handling safety regulation, and it is the same shift in perspective this book is attempting to instill more broadly across every domain where energy accumulates silently before it, eventually, does not.

The next chapter moves to a category of silent energy transformation governed by entirely different physical principles: not chemical bond rearrangement, not nuclear decay, not biological metabolism, but the physics of phase transitions, the surprisingly dramatic energy changes hidden inside the everyday process of melting, freezing, boiling, and condensing.

CHAPTER NINE

Frozen Fire

Fill a plastic bottle with pure, still water. Chill it slowly and carefully in a freezer set just below the standard freezing point, taking care not to disturb it. Under the right conditions, something strange happens: the water does not freeze. It remains entirely liquid, sitting quietly at a temperature several degrees below zero Celsius, defying what every school science lesson taught you about when water is supposed to turn to ice. This state, called supercooling, is entirely real, entirely well documented in physics, and entirely unstable. The moment that bottle is disturbed, tapped sharply on a countertop, or has a single ice crystal or particle of dust introduced into it, the water inside can transform from liquid to solid ice almost instantaneously, sometimes visibly crystallizing in a wave that spreads through the entire container within a fraction of a second, releasing a genuine, measurable burst of heat as it does so.

This chapter is about that release, and about the broader category of silent energy transformation it represents: not the rearrangement of chemical bonds examined in earlier chapters, and not the nuclear rearrangement examined in chapter four, but the physics of phase transitions, the changes between solid, liquid, and gas that quietly govern an enormous range of natural and engineered phenomena, and that hide a genuinely surprising quantity of energy inside processes most people consider entirely mundane.

The Energy Hiding Inside Melting Ice

When you place an ice cube in a warm drink, most people intuitively understand that the ice absorbs heat from the surrounding liquid, cooling the drink. Far fewer people appreciate just how much heat that process actually absorbs, and why. As ice melts, the water molecules within it are not simply speeding up in the way molecules do when a substance's temperature rises. They are, instead, breaking free of the rigid crystalline lattice structure that defines ice as a solid, a structural transformation that requires energy entirely separate from, and considerably larger than, the energy needed to simply raise the temperature of an equivalent mass of already liquid water by a single degree.

This energy, required to change a substance's physical state without changing its temperature, is called latent heat, and for water specifically, the numbers involved are unusually large compared to most other common substances. Melting one gram of ice at zero degrees Celsius into one gram of liquid water at the same temperature requires approximately 334 joules of energy, according to standard physics reference values cited consistently across scientific literature. This might sound like a modest figure until you compare it directly to the energy required simply to heat that same gram of already liquid water. The specific heat capacity of water, meaning the energy required to raise one gram of liquid water by one degree Celsius, is approximately 4.18 joules. This means the energy required merely to melt a gram of ice, without changing its temperature at all, is equivalent to the energy that would be required to heat that same mass of already liquid water by roughly eighty degrees Celsius, nearly the entire distance from freezing to boiling.

This is precisely why ice is such a remarkably effective cooling agent, far more effective, gram for gram, than an equivalent mass of merely cold water would be. When ice cools a drink, it is not simply warming up from a cold solid to a slightly less cold solid. It is absorbing an enormous quantity of latent heat purely in the act of transforming from solid to liquid, heat drawn directly from the surrounding drink, and only after that transformation is complete does the resulting liquid water begin absorbing additional heat in the more conventional, temperature raising sense. This exact physical principle, incidentally, is precisely what powered the ice based cooling system described in chapter seven, where engineers building Hoover Dam relied on a dedicated refrigeration plant to produce roughly a thousand pounds of ice daily, understanding intuitively, even if not always in these precise thermodynamic terms, that ice represented a far more concentrated and efficient way to extract heat from curing concrete than an equivalent mass of merely cold water circulating through the same pipe network would have provided.

Boiling Hides an Even Larger Number

If the latent heat of fusion, the energy involved in melting or freezing, is already substantial, the latent heat of vaporization, the energy involved in boiling or condensing, is considerably larger still. Converting one gram of water at its boiling point into steam at that same temperature requires approximately 2,230 joules of energy, according to standard reference figures used throughout physics and engineering education, nearly seven times the energy required to melt an equivalent mass of ice.

This enormous figure explains a great deal about the physical world that might otherwise seem mysterious. It explains why steam burns are frequently far more severe than burns from water at the same one hundred degree Celsius temperature, because steam condensing on skin releases that entire enormous latent heat of vaporization directly into the tissue it contacts, delivering a quantity of thermal energy dramatically larger than an equivalent mass of already boiling liquid water would transfer on contact. It explains why steam has historically been, and remains today, one of the most important working fluids in power generation, because steam turbines exploit precisely this enormous stored energy, extracting useful mechanical work as steam condenses back into liquid water inside a turbine system, a principle underlying the large majority of the world's electricity generation, whether the initial heat source is burning coal, natural gas, or the nuclear fission reactions examined briefly in chapter four. And it explains why sweating is such a remarkably effective biological cooling mechanism, a topic examined in greater depth in the next chapter, because the evaporation of sweat from skin draws precisely this same large latent heat of vaporization directly from the body, providing a cooling effect far more powerful than the modest temperature difference between skin and air alone could ever achieve.

None of this energy transfer, in any of these examples, involves flame, chemical combustion, or nuclear reaction of any kind. It is purely physical, governed entirely by the energy required to rearrange the spacing and bonding pattern between molecules as a substance shifts between solid, liquid, and gas phases, and it proceeds, in the overwhelming majority of everyday circumstances, in complete silence, entirely invisible to casual observation, exactly like every other silent explosion examined throughout this book.

The Strange Case of Supercooled Water

Return now to the supercooled water bottle described at the start of this chapter, because it represents one of the more genuinely striking, and genuinely useful to understand, demonstrations of latent heat in action. Under carefully controlled conditions, free of the microscopic impurities, surface imperfections, or physical disturbances that would normally provide a starting point for ice crystal formation, pure water can remain entirely liquid at temperatures well below its nominal freezing point, a metastable state that physicists describe using the term supercooling.

This metastable state is inherently unstable, and the mechanism that ends it illustrates the underlying physics of phase transitions with unusual clarity. Ice crystal formation requires what physicists call a nucleation site, a microscopic starting point, perhaps a tiny air bubble, a speck of dust, a scratch on the container's interior surface, or simply a sufficiently forceful physical disturbance, around which the first ice crystal can begin forming. In carefully prepared supercooled water, free of these nucleation sites, the liquid state can persist metastably even below the normal freezing point, because there is, quite literally, no starting point available for the crystallization process to begin.

Introduce a nucleation site, however, even something as simple as a single ice chip dropped into the bottle, or a sudden sharp tap against a hard surface, and the situation changes instantly. Ice crystals begin forming at that nucleation point, and because the surrounding water is already below its normal freezing temperature, the crystallization process propagates through the remaining liquid at remarkable speed, often visibly sweeping through an entire bottle of water within a matter of seconds. Each portion of water that crystallizes releases its latent heat of fusion, exactly as described earlier in this chapter, approximately 334 joules per gram, and this released heat actually helps explain a subtlety that puzzles many people encountering this phenomenon for the first time: supercooled water does not necessarily freeze completely and instantly the moment nucleation begins. Often, only a portion of the total water mass crystallizes rapidly, with the latent heat released by that initial crystallization warming the remaining liquid water back up to almost exactly zero degrees Celsius, the normal equilibrium freezing point, at which point the remaining liquid water settles into an ordinary, gradual freezing process rather than continuing the dramatic, rapid crystallization that characterized the initial nucleation event. This is a genuine, calculable energy balance, precisely the kind of quantitative relationship physicists studying supercooled water phenomena have worked through in detail, and it demonstrates, in miniature, exactly the same principle of energy release and reaction rate discussed throughout this book: a system sitting in an unstable, energy laden state, quietly, until a trigger, in this case a single nucleation event, releases that stored energy rapidly and dramatically.

Phase Transitions in Nature and Engineering

Beyond the party trick appeal of supercooled water, phase transition energy plays a genuinely significant, often underappreciated role across numerous natural and engineered systems, and recognizing this role helps complete the picture this book has been building throughout its earlier chapters.

Consider frost protection in agriculture, a practice that directly exploits the latent heat of fusion to protect crops from freezing damage. Citrus growers and other farmers vulnerable to frost damage have long used a technique involving spraying crops with water during freezing conditions, a practice that seems, on its surface, deeply counterintuitive. Why would deliberately adding water to plants already at risk of freezing help protect them from cold damage. The answer lies entirely in latent heat. As the sprayed water on the plant surface freezes, it releases its latent heat of fusion directly into the immediate surrounding environment, specifically the plant tissue beneath the forming ice layer, and this released heat can keep the plant tissue itself from dropping below the critical temperature at which genuine cellular damage occurs, even while a layer of ice visibly forms on its exterior surface. The ice coating, counterintuitively, is protecting the plant precisely because its formation is releasing real, measurable thermal energy exactly where the plant needs it most.

Consider also the vast quantity of thermal energy involved in the seasonal freezing and thawing of polar ice, sea ice, and glacial systems, a phase transition energy budget so enormous that it plays a genuinely significant role in global and regional climate systems, moderating temperature swings in ways that climate scientists account for explicitly in their models of Earth's overall energy balance. The latent heat released as ocean water freezes into sea ice each polar winter, and absorbed again as that ice melts each polar summer, represents an energy exchange on a scale that dwarfs nearly every other silent explosion examined in this book, distributed across millions of square kilometers of ocean surface, entirely without producing a single sound audible to any observer, and entirely without requiring a single chemical bond to break or reform anywhere in the process.

Why Phase Transitions Belong in a Book About Explosions

It might seem, at first consideration, unusual to include melting ice and boiling water in a book primarily concerned with dramatic, energetic phenomena like rusting steel, radioactive decay, and battery thermal runaway. But the underlying logic connecting this chapter to the rest of the book is exactly the same logic established in chapter two: energy release does not require chemical bonds breaking, and it does not require an audible signature. It requires only a system transitioning from a higher energy configuration to a lower one, or in the case of melting and boiling, absorbing energy to transition from a lower energy, more ordered configuration to a higher energy, less ordered one. The physics governing exactly how much energy is involved in these transitions is every bit as precise, every bit as measurable, and every bit as significant to real world engineering and natural systems as the chemistry governing combustion or the nuclear physics governing radioactive decay.

What phase transitions add to the picture this book has been building is a reminder that silent energy release is not confined to chemistry and nuclear physics alone. It extends into the most basic physical transformations imaginable, the freezing of water, the boiling of a kettle, the quiet, continuous exchange of latent heat happening at every ice cube melting in every drink poured anywhere in the world at this exact moment, each one absorbing a small, precisely calculable, entirely silent burst of energy as it makes the transition from solid to liquid, a transformation that would strike nobody as remarkable, but that obeys the exact same fundamental thermodynamic principles as every dramatic, headline making silent explosion examined elsewhere in these pages.

The next chapter turns inward, from ice cubes and steam turbines to the human body itself, and examines the remarkable, precisely regulated silent chemistry happening inside every living cell in your body at this very moment, chemistry responsible for keeping your body at a stable temperature, for generating the energy that powers every thought and movement, and for occasionally, under specific and fascinating circumstances, generating dramatically more heat than ordinary metabolism requires.

CHAPTER TEN

The Body's Hidden Reactions

Newborn infants cannot shiver in any meaningful, effective way. Their muscle mass is too small, their motor control too undeveloped, to generate significant heat through the rapid, involuntary muscle contractions that older children and adults rely on when exposed to cold. And yet healthy newborns, remarkably, do not simply lose the battle against cold exposure the way this deficiency might suggest they should. They possess a specialized tissue, largely absent or greatly diminished in most adult humans until relatively recent research revealed otherwise, called brown adipose tissue, or brown fat, which does something no other tissue in the human body does in quite the same way: it burns fuel specifically to produce heat, deliberately sacrificing the usual efficiency of cellular energy production in favor of pure, direct thermal output.

This chapter is about that tissue, and more broadly about the entire silent, continuous chemical furnace that keeps your body at a stable ninety eight point six degrees Fahrenheit, hour after hour, day after day, entirely without your conscious awareness, powered by chemistry every bit as real, every bit as measurable, and every bit as exothermic as the reactions examined throughout the rest of this book.

The Currency of Cellular Energy

Every cell in your body runs on a molecule called adenosine triphosphate, universally abbreviated as ATP, which functions as the immediate, usable currency of cellular energy, the biological equivalent of cash in your pocket, as opposed to the larger, less immediately accessible reserves of chemical energy stored in fat and glycogen, which function more like money sitting in a savings account, requiring conversion before it becomes directly usable.

ATP stores energy in the chemical bonds connecting its three phosphate groups, and when a cell needs energy for any purpose, muscle contraction, nerve signal transmission, protein synthesis, active transport of molecules across a cell membrane, it breaks one of these phosphate bonds, releasing energy directly usable for that specific cellular task, and converting ATP into adenosine diphosphate, or ADP, in the process. This entire cycle, ATP breaking down into ADP to release usable energy, and ADP being rebuilt back into ATP using energy harvested from food, happens on an almost unimaginable scale continuously throughout your body. The human body does not maintain a large standing reserve of ATP at any given moment. Instead, it continuously recycles a comparatively small pool of ATP molecules, breaking them down and rebuilding them at a staggering rate, estimated by biochemists to total, over the course of a single day for an average adult at rest, a mass of ATP roughly equivalent to that person's own body weight, entirely through this continuous cycle of breakdown and reconstruction rather than through the synthesis of that quantity from scratch.

The process by which cells rebuild ATP, called cellular respiration, is fundamentally an oxidation reaction, chemically related, at the most basic level, to the combustion reactions examined throughout this book, though vastly more controlled, more efficient, and more precisely regulated than an open flame. Glucose, derived from the food you eat, is systematically broken down through a sequence of enzyme catalyzed reactions, ultimately combining with oxygen you breathe in to produce carbon dioxide, water, and usable energy captured in the form of newly synthesized ATP. The overall chemistry bears a real resemblance to the combustion of a simple hydrocarbon fuel, releasing energy through oxidation, but cellular respiration accomplishes this oxidation through dozens of carefully sequenced, enzyme mediated intermediate steps, each releasing a small, precisely controlled quantity of energy, rather than through the single, rapid, high temperature reaction characteristic of an open flame. This is precisely why your body can extract usable energy from food at a gentle ninety eight point six degrees Fahrenheit, a temperature far too low to sustain open combustion of the same fuel molecules, using instead a sophisticated cascade of enzyme catalysts, discussed conceptually in chapter two, that lower the effective activation energy barrier for each individual step in the overall reaction sequence.

Why Some Heat Production Is Deliberately Wasteful

Ordinarily, cellular respiration is a remarkably efficient process, in the specific sense that a large proportion of the chemical energy extracted from food is successfully captured in the form of newly synthesized ATP, available for the cell's actual functional needs, rather than simply dissipated as heat. But efficiency, in this specific biochemical sense, is not always the goal, and brown adipose tissue represents the clearest example in human physiology of cellular machinery deliberately engineered, through millions of years of evolutionary refinement, to sacrifice that efficiency entirely in favor of maximizing pure heat output instead.

The mechanism responsible, extensively documented in physiological research, centers on a specialized protein called uncoupling protein one, commonly abbreviated as UCP1, found specifically within the mitochondria of brown fat cells. In ordinary cellular respiration, occurring in essentially every cell type throughout the body, the biochemical machinery of the mitochondria creates a gradient of hydrogen ions across an internal membrane, and this gradient is then carefully harnessed by a specific enzyme complex to synthesize ATP, in a tightly coupled process where the energy released by nutrient oxidation is directly channeled into ATP production. UCP1, as researchers studying brown fat physiology have documented in detail, functions as a kind of deliberate short circuit in this process, providing an alternative pathway that allows hydrogen ions to flow back across the mitochondrial membrane without passing through the ATP synthesizing machinery at all. This means the energy that would ordinarily have been captured as ATP is instead released directly as heat, essentially converting the mitochondria of brown fat cells into small, dedicated, biological furnaces, deliberately uncoupling fuel oxidation from useful energy capture, specifically in order to maximize thermal output instead.

The physiological significance of this mechanism is genuinely substantial, not merely a curious biochemical footnote. Research examining brown adipose tissue activity in small mammals living in cold environments has found that nearly half of their total energy metabolism can be attributed specifically to brown fat thermogenesis under cold stress conditions, according to physiological research published in peer reviewed literature. In humans, while brown fat's proportional contribution to total adult energy expenditure is smaller than in small mammals adapted to extreme cold, researchers using thermal imaging and metabolic tracking techniques have estimated that as little as fifty grams of active brown adipose tissue is capable of burning approximately two hundred and fifty kilocalories daily, according to figures cited in research examining brown fat thermogenesis at the individual cell level, a genuinely significant contribution to total daily energy expenditure from a comparatively small mass of specialized tissue, and a finding that has driven substantial pharmaceutical research interest in the possibility of therapeutically activating brown fat as a treatment strategy for obesity and related metabolic disorders.

Fever as Deliberate, Controlled Overheating

Brown fat thermogenesis is not limited to protecting against cold exposure. Research into the physiological control of body temperature has documented that the same underlying thermogenic pathway, involving brown adipose tissue activation, plays a direct, well characterized role in fever, the controlled elevation of body temperature that occurs during infection. This finding reframes fever in a way that many people, understandably, never consider: it is not a passive, unwanted symptom of illness, a malfunction to be immediately suppressed, but rather an active, deliberately generated physiological response, driven by real, measurable thermogenic chemistry, that the immune system employs specifically because elevated body temperature genuinely impairs the viability of many pathogens and enhances the effectiveness of various immune cell responses, according to physiological research examining the mechanisms connecting brown fat activity to the febrile response.

This is a striking illustration of the same principle established throughout this book applied directly to your own body: a genuinely significant, deliberately controlled release of chemical energy as heat, triggered by a specific biological signal, in this case pyrogenic mediators released by the immune system in response to detected infection, proceeding entirely through ordinary cellular chemistry, without flame, without any external heat source, and without any sound whatsoever, and yet capable of measurably raising your entire body's temperature by one, two, or occasionally several degrees, a genuinely substantial thermal effect achieved purely through the biochemistry described in this chapter.

The Everyday Furnace You Never Notice

Even setting aside the specialized case of brown fat thermogenesis, the ordinary metabolic activity of a resting adult human body generates a genuinely substantial, continuous thermal output, entirely as an unavoidable byproduct of the cellular respiration processes described earlier in this chapter, occurring in essentially every cell throughout the body simultaneously. This baseline heat production, commonly referenced by physiologists using the concept of basal metabolic rate, is precisely why a room full of people feels noticeably warmer than an empty room, why sleeping bags and thermal blankets work primarily by trapping the heat your own body is continuously generating rather than by actively producing warmth themselves, and why maintaining stable body temperature, a physiological process called thermoregulation, represents one of the most continuously active, energetically significant tasks your body performs at every moment of your life, whether you are consciously aware of it or not.

This continuous thermal output is not, strictly speaking, waste in the way heat lost from an inefficient machine might be considered waste. A significant portion of it is genuinely functional, maintaining the precise temperature range within which the enzymes governing your metabolism, discussed earlier in this chapter, and referenced conceptually in chapter two regarding the role catalysts play in accelerating chemical reactions, function optimally. Enzymes are remarkably temperature sensitive, and the elaborate thermoregulatory systems your body maintains, involving blood vessel dilation and constriction, sweating, shivering, and the brown fat thermogenesis described above, exist specifically to keep your internal chemistry operating within the comparatively narrow temperature window where your enzymes function most effectively, a window centered closely around that familiar ninety eight point six degree Fahrenheit figure most people learn as children without ever fully appreciating just how much continuous, silent, genuinely exothermic chemistry is required to maintain it.

Evaporative Cooling and the Physics From the Previous Chapter

The thermoregulatory system responsible for preventing your body from overheating, meanwhile, directly exploits the latent heat physics established in the previous chapter. Sweating cools the body specifically because the evaporation of sweat from skin requires drawing latent heat of vaporization, the same physical quantity discussed in chapter nine, approximately 2,230 joules for every gram of water evaporated, directly from the skin and the blood circulating immediately beneath it. This is precisely why sweating in genuinely dry conditions, where evaporation proceeds efficiently, provides dramatically more effective cooling than sweating in humid conditions, where the surrounding air is already close to saturated with water vapor and evaporation consequently proceeds far more slowly, leaving sweat sitting on the skin without providing its intended cooling benefit nearly as effectively.

This connection between chapters nine and ten is worth highlighting explicitly, because it demonstrates something important about how the various silent explosions examined throughout this book are not, in fact, entirely separate, unrelated phenomena, but rather different expressions of a shared underlying set of physical principles, chemistry, thermodynamics, and phase transition physics, appearing again and again across radically different contexts, from a rusting bridge, to a curing dam, to the sweat cooling your skin after a run on a hot day.

A Body Full of Silent Explosions

Step back from the specific biochemical detail examined throughout this chapter, and a genuinely remarkable picture emerges. Your body, at this exact moment, is running an enormous, continuous, tightly regulated network of oxidation reactions, chemically related to the combustion reactions examined in earlier chapters, converting the chemical energy stored in the food you have eaten into usable cellular energy and, as an inevitable and often deliberately amplified byproduct, genuine thermal heat. This entire process proceeds at a temperature low enough that open flame combustion of the same fuel molecules would be entirely impossible, made possible instead through an elaborate, evolutionarily refined system of enzyme catalysts that lower the effective activation energy for each individual reaction step, precisely the mechanism described conceptually back in chapter two. Specialized tissue exists specifically to deliberately sacrifice metabolic efficiency in favor of pure heat generation, whether protecting a newborn infant from cold exposure or contributing to the deliberately elevated body temperature of a fever fighting off infection. And the same latent heat physics governing melting ice and boiling water, examined in the previous chapter, governs the evaporative cooling mechanism your body relies on to prevent overheating during exertion or in hot conditions.

None of this chemistry makes a sound. None of it produces visible flame. And yet, measured purely in terms of the total quantity of chemical energy processed, converted, and ultimately released as heat over the course of a single day, an ordinary human body represents a genuinely significant, continuously operating chemical reactor, quietly converting stored chemical energy into usable cellular work and thermal output, second by second, entirely within the range of ordinary biological temperature, entirely without your conscious involvement, for every single moment of your life from the instant you are born until the moment that final chemistry finally stops.

The next chapter leaves the human body behind and turns to a much larger scale entirely: the geological processes deep within the Earth itself, where exothermic mineral reactions, radioactive decay heat within the planet's interior, examined earlier in chapter four but now considered at planetary scale, and the immense, continuous energy budget of geothermal systems combine to make the ground beneath your feet a far more energetically active place than most people ever pause to consider.

CHAPTER ELEVEN

Geological Giants

Beneath your feet, right now, regardless of where on Earth you happen to be reading this sentence, the planet is releasing energy at a rate scientists have measured, through decades of careful geophysical study, at approximately forty four to forty seven terawatts, flowing continuously outward from Earth's interior toward the surface, according to figures cited across multiple peer reviewed geophysics studies examining Earth's total heat budget. A terawatt is a trillion watts. Forty seven of them, released continuously, every second, day and night, winter and summer, entirely independent of weather, sunlight, or season, in a process so vast in scale and so gradual in its surface expression that most people spend their entire lives walking across it without ever considering that the ground itself is, in a very real sense, a colossal, slow burning engine.

This chapter examines where that staggering quantity of energy comes from, and it turns out the answer connects directly back to two mechanisms already established earlier in this book: the radioactive decay examined in chapter four, now considered at a scale spanning the entire planet, and a category of chemical reaction, occurring deep within Earth's crust and involving some of the most abundant minerals on the planet, that few people outside of specialized geology have ever heard of, despite its potential relevance to some of the biggest open questions in planetary science.

The Radioactive Furnace, Scaled to a Planet

Chapter four examined how a small canister of plutonium 238 can quietly power a spacecraft for decades through nothing more than the steady heat released by nuclear decay. Earth's interior runs on precisely the same fundamental physics, just distributed across a planetary volume and involving different, naturally occurring radioactive isotopes, primarily uranium 238, uranium 235, thorium 232, and potassium 40, present in genuinely trace concentrations throughout Earth's crust and mantle, but multiplied across a mass so enormous that even those trace concentrations add up to a substantial continuous energy source.

Geophysicists refer to this contribution as radiogenic heat, and determining precisely how much of Earth's total heat budget it accounts for has proven to be a genuinely difficult scientific problem, requiring remarkably creative measurement techniques. Because these radioactive isotopes are distributed throughout Earth's interior, largely inaccessible to direct sampling given that the deepest boreholes humans have ever drilled reach only a small fraction of the distance to Earth's core, scientists have turned to an unusual and elegant solution: detecting geoneutrinos, nearly massless subatomic particles produced directly by the radioactive decay processes happening deep within the planet, which pass through solid rock largely unimpeded and can be detected using specialized underground particle physics detectors originally built for other purposes. Research using this geoneutrino detection technique has estimated that radioactive decay accounts for roughly forty percent of Earth's total surface heat flow, according to figures published in physics research examining directional neutrino measurements of Earth's interior, with the remaining heat attributed to primordial energy left over from the planet's initial formation and the ongoing gradual solidification of Earth's molten core, a process that itself continues releasing gravitational and frictional energy as it very slowly proceeds.

This radiogenic heat contribution is not merely a scientific curiosity. It plays a direct, causal role in some of the largest scale geological processes shaping the planet's surface. The heat driving mantle convection, the slow, continuous churning motion of Earth's mantle that ultimately powers plate tectonics, the process responsible for continental drift, mountain building, earthquake activity, and volcanic eruptions, derives substantially from this same radioactive decay process, according to geological research examining Earth's thermal history. In an almost poetic sense, the same fundamental physical process responsible for the quiet warmth radiating from a plutonium powered spacecraft battery, drifting silently through interstellar space, is also responsible, operating at a vastly larger scale over geological timescales measured in hundreds of millions of years, for the formation of the Himalayas, the eruption of Mount Vesuvius, and the ongoing slow separation of the African and Arabian tectonic plates.

A Chemical Reaction That Cooks Rock From the Inside

Radiogenic heat, however, is not the only silent, large scale exothermic process occurring within Earth's crust. There exists a genuinely fascinating chemical reaction, called serpentinization, in which certain iron and magnesium rich minerals, most notably olivine, one of the most abundant minerals in Earth's upper mantle, react chemically with water that has infiltrated the crust, transforming into a family of minerals called serpentine, and releasing a substantial quantity of heat in the process.

Research published in planetary science literature examining this reaction has documented its enthalpy, the total heat released per unit mass of reacting material, at approximately four hundred and fifty thousand joules per kilogram, a genuinely significant figure that places serpentinization among the more energetic exothermic reactions examined anywhere in this book, on a per kilogram basis. What makes this reaction particularly significant to planetary scientists is not merely its magnitude, but its ubiquity and its potential role in processes far beyond ordinary Earth geology. Because the reaction requires only two comparatively common ingredients, olivine, an abundant rock forming mineral found throughout Earth's mantle and in many meteorites, and liquid water, researchers studying planetary formation have proposed that serpentinization may have played a genuinely significant role in heating small icy bodies throughout the early solar system, including planetesimals, the small rocky and icy building blocks from which planets eventually formed, and potentially even objects in the outer solar system's Kuiper Belt, according to research published in the Monthly Notices of the Royal Astronomical Society examining the thermal evolution of planetesimals.

This is a striking extension of the core argument running throughout this entire book, taken to a genuinely cosmic scale. A silent, entirely non explosive, non luminous chemical reaction between rock and water, the same fundamental category of chemistry, oxidation and hydration reactions releasing energy through molecular rearrangement, examined in chapter three's discussion of rust and chapter seven's discussion of cement hydration, may have contributed meaningfully to heating small, icy planetary building blocks scattered throughout the early solar system billions of years ago, an entirely invisible process, occurring far from any observer, in the frigid vacuum of space, that nonetheless played a genuine, measurable role in shaping the geological and chemical evolution of small bodies that, in some cases, still orbit the sun today, largely unchanged since that early period of solar system history.

Why Earth's Crust Feels Cool While Its Interior Radiates Heat

A natural question arises from the figures presented in this chapter: if Earth is genuinely releasing forty seven terawatts of heat continuously, why does the ground beneath your feet feel, in most locations, entirely ordinary, neither noticeably warm nor obviously connected to any dramatic underlying energy source. The answer illustrates a principle established repeatedly throughout this book, and worth restating explicitly here: total energy and power density, meaning energy released per unit area or per unit time in a specific location, are entirely different measurements, and a genuinely enormous total energy budget can still translate into an imperceptibly small effect at any single point on the surface, simply because that energy is distributed across an enormous total area.

Earth's total surface area is approximately five hundred and ten million square kilometers. Distribute forty seven terawatts of continuous heat flow evenly across that entire surface, and the resulting average heat flux per unit area works out to a genuinely modest figure, sufficient to be measured with sensitive scientific instruments but far too small to be perceptible to ordinary human senses standing on an average patch of ground anywhere on the planet. This is precisely why, outside of specific geologically active regions, geysers, hot springs, active volcanic zones, and areas of unusually thin crust where the underlying heat becomes locally concentrated enough to reach the surface in an obviously detectable form, most of the planet's surface gives no outward sign whatsoever of the staggering total quantity of energy continuously flowing outward from its interior.

Geothermal energy production, an increasingly significant contributor to renewable electricity generation in various countries around the world, exists specifically to exploit those particular locations where geological conditions concentrate Earth's internal heat closer to the surface than the global average, allowing engineers to tap directly into a portion of this otherwise diffuse, imperceptible energy flow and convert it into genuinely useful electrical power. Iceland, positioned directly atop the Mid Atlantic Ridge, where tectonic plates are actively separating and allowing Earth's internal heat unusually direct access to the surface, has become a particularly well known example of a nation deriving a substantial proportion of its electricity and heating needs directly from this geothermal resource, according to figures widely cited in energy research examining renewable geothermal power generation.

Volcanoes as the Loud Exception That Proves the Rule

It is worth pausing to note explicitly that this chapter's subject matter sits in an interesting position relative to the broader argument of this book. Most of the geological energy release examined so far, radiogenic decay heat and serpentinization chemistry, proceeds in complete silence, distributed gradually across enormous volumes of rock and enormous spans of geological time. Volcanic eruptions represent the dramatic, loud exception, the moment when accumulated geological energy, pressure, and molten material finally breach the surface in a sudden, sometimes catastrophic release, entirely analogous to the deflagration to detonation transition examined in chapter eight's discussion of grain dust explosions, or the thermal runaway cascade examined in chapter five's discussion of lithium battery failure.

The underlying pattern is, once again, identical to nearly every silent explosion examined throughout this book. Pressure and thermal energy accumulate gradually, silently, over a period that can stretch from years to centuries depending on the specific volcanic system involved, entirely invisible to surface observation except for specialized geological monitoring instruments capable of detecting the subtle ground deformation, gas emission changes, and minor seismic activity that often precede major eruptions. Then, upon crossing a specific physical threshold, when accumulated pressure finally exceeds the structural strength of the overlying rock, that gradually accumulated energy releases suddenly, dramatically, and in the case of the largest volcanic eruptions in recorded history, with consequences felt across the entire planet, from measurable global temperature drops caused by ash and aerosols injected into the upper atmosphere, to significant disruption of aviation and agriculture across enormous geographic regions far from the eruption site itself.

The Ground Beneath a New Understanding

This chapter closes the geographic scope of this book's argument at its widest possible scale, moving from the microscopic chemistry of a rusting nail, through the human body, and finally to the entire planet itself, understood as a single, continuously operating thermal system, powered substantially by the same radioactive decay physics that powers a spacecraft's electrical systems, and supplemented by exothermic mineral chemistry that may have shaped the earliest chapters of our solar system's geological history. Every step of that scale, from nail to planet, obeys the exact same fundamental thermodynamic and chemical principles established at the very start of this book: energy release does not require sound, does not require flame, and does not require any particular scale, small or large, to be genuine, measurable, and in many cases, genuinely consequential to the world as we experience it.

The remaining chapters of this book turn from the natural world back toward human technology and engineering, examining in closer detail the specific dangers, and the specific engineering solutions, associated with two of the more consequential silent explosion risks in modern life: the lithium battery chemistry that powers much of contemporary consumer electronics and transportation, considered here in greater technical depth than chapter five allowed, and the industrial hazards associated with reactive metals and materials handled at scale in modern manufacturing, before this book concludes with a broader examination of how engineers across every domain examined so far have learned, often at considerable cost, to take silence seriously as a warning sign rather than dismissing it as evidence of safety.

CHAPTER TWELVE

The Silent Bomb in Your Pocket

According to the most recent analysis published by the United States National Highway Traffic Safety Administration, it takes approximately seven point two million miles of electric vehicle travel to produce a single vehicle fire. The comparable figure for gasoline powered vehicles is approximately eighteen thousand miles, meaning gasoline vehicles catch fire, per mile driven, at a rate roughly four hundred times higher than electric vehicles. Data compiled from multiple national fire agencies tells a consistent story: battery electric vehicles experience approximately twenty five fires per one hundred thousand vehicles sold, compared to roughly one thousand five hundred per one hundred thousand for gasoline vehicles, according to figures widely cited across recent automotive safety research, with Swedish national fire statistics showing an even starker contrast, three point eight fires per one hundred thousand electric and hybrid vehicles compared to sixty eight per one hundred thousand for combustion engine vehicles.

This chapter opens with these statistics deliberately, because they complicate a popular narrative in a way that matters enormously to the argument this book has been building. Lithium batteries, examined already in chapter five through the specific lens of the Boeing 787 and Samsung

Note 7 incidents, have earned a reputation in public consciousness as unusually dangerous, a reputation reinforced by dramatic viral video footage and vivid, visually striking failure modes. And yet the data consistently shows that, measured across large populations of vehicles and devices, lithium battery fires are considerably rarer than fires involving the flammable liquid fuel technology that has powered vehicles for over a century, precisely the kind of technology that produces such familiar, unremarkable combustion that almost nobody thinks to describe gasoline as dangerous in ordinary daily conversation.

This chapter goes deeper into the underlying chemistry of lithium batteries than chapter five's incident focused examination allowed, in order to understand precisely why these devices, despite their genuinely dramatic and occasionally newsworthy failure mode, remain, statistically, one of the safer widely deployed energy storage technologies available today, and why the specific engineering response to their risks offers one of the clearest, most instructive examples in this entire book of how silent hazards get systematically tamed once they are properly understood.

Why Lithium Specifically

To understand lithium battery chemistry, it helps to understand why lithium, specifically, became the element of choice for high performance rechargeable batteries in the first place, a choice with direct consequences for both the technology's remarkable energy density and its particular safety considerations.

Lithium sits at the very top left corner of the periodic table, the lightest metal that exists in solid form under ordinary conditions, and this position confers a specific, extremely useful electrochemical property: a high electrochemical potential relative to its atomic mass, meaning lithium atoms can, in principle, store and release a comparatively large quantity of electrical energy relative to how much they weigh. This property, sometimes summarized by battery engineers as high energy density, is precisely why lithium based batteries have become the dominant technology across smartphones, laptops, and electric vehicles, applications where minimizing weight and maximizing stored energy in a compact space both matter enormously.

But this same property that makes lithium so useful also explains why lithium battery chemistry carries the specific risk profile examined in chapter five. A material capable of storing a large quantity of electrical energy in a small mass is, by definition, capable of releasing that same large quantity of energy quickly if the normal controlled discharge pathway is disrupted. Industry research examining lithium battery safety has documented explicitly that lithium's high electrochemical potential, precisely the property responsible for its usefulness as an anode material, simultaneously makes it more susceptible to the kind of thermal runaway failure examined in detail in chapter five, a genuine engineering tradeoff between performance and inherent safety margin that battery engineers have spent decades working to manage through cell design, manufacturing quality control, and increasingly sophisticated battery management systems rather than through abandoning lithium chemistry altogether.

The Real Numbers Behind the Danger

It is worth being specific and quantitative about exactly how the different causes of electric vehicle fires break down, because the picture that emerges is considerably more nuanced than the simple narrative of dangerous batteries spontaneously combusting that dominates casual public discussion. Data compiled from the Australian EV FireSafe research program, tracking incidents from 2010 through 2023, found that collision related causes accounted for fifty one percent of electric vehicle fire incidents, meaning the majority of fires involved genuine physical damage to the vehicle and its battery pack from an accident, rather than a spontaneous failure occurring during normal operation. Norwegian data covering 2022 specifically attributed thermal runaway to twenty three percent of electric vehicle fire incidents in that country, according to figures compiled in recent automotive fire safety research, a meaningful but clearly minority share of total incidents once collision damage, charging equipment faults, and other external causes are properly accounted for separately.

This breakdown matters because it reframes the underlying engineering challenge accurately. Lithium battery packs, when left entirely undisturbed and charged and discharged within their designed operating parameters, are, based on the accumulated real world data from millions of vehicles now in service worldwide, genuinely reliable and safe, consistent with the fundamental four stage thermal runaway mechanism described in chapter five, which requires a specific triggering event, an internal short circuit from a manufacturing defect, physical damage from a crash or puncture, or abuse conditions like severe overcharging, none of which occur spontaneously during ordinary, undisturbed operation. The genuine engineering challenge, and the genuine remaining safety frontier for this technology, centers specifically on hardening battery packs against exactly these triggering events: collision damage, manufacturing defects, and charging system faults, rather than on some inherent, unavoidable instability in lithium chemistry itself.

The Engineering Response, Layer by Layer

Modern lithium battery pack design, particularly in electric vehicles where the sheer number of individual cells involved, often numbering in the thousands within a single vehicle's battery pack, makes even a very small per cell failure probability a meaningful aggregate concern, has evolved into a genuinely sophisticated, multi layered defense system, and walking through these layers illustrates exactly how engineers have systematically addressed each stage of the thermal runaway process described in chapter five.

At the cell level, manufacturing quality control has improved dramatically since the era of the Boeing 787 and Samsung Note 7 incidents, driven directly by the root cause investigations into precisely those failures. Modern lithium cell manufacturing incorporates increasingly rigorous inspection processes specifically designed to detect the kind of internal manufacturing defects, crimped edges, rough electrode surfaces, foreign debris contamination, that were identified as root causes in both of those earlier incidents, using techniques ranging from advanced X ray imaging of individual cells to statistical process control methods borrowed from semiconductor manufacturing, an industry with its own long history of needing to detect microscopic defects in mass produced components.

At the pack level, modern battery management systems continuously monitor individual cell voltage, temperature, and in more sophisticated designs, internal impedance, across every single cell within a larger battery pack, in real time, throughout both charging and discharging. This monitoring is specifically designed to catch the earliest possible indicators of a cell beginning to deviate from normal behavior, potentially identifying a cell entering the very first stage of thermal runaway described in chapter five, well before that cell's temperature rise becomes severe enough to threaten neighboring cells, and modern systems are capable of electrically isolating a suspect cell from the rest of the pack automatically, preventing the kind of cascading, pack wide failure that made the Note 7 incidents so dramatic and so costly.

At the structural level, modern electric vehicle battery packs increasingly incorporate physical barriers, specifically engineered to prevent thermal propagation between individual cells or cell modules, meaning that even if a single cell does enter thermal runaway despite the preceding layers of defense, the resulting heat and any ejected material remains physically contained, rather than spreading to trigger a cascading failure across the entire pack. This design philosophy, sometimes described in automotive engineering literature as containing rather than preventing worst case failures, reflects a mature, layered approach to risk management that accepts individual cell failures will occasionally occur despite best efforts at prevention, while ensuring those individual failures cannot cascade into the kind of large scale, structure threatening event that characterized the earlier aviation and consumer electronics incidents examined in chapter five.

Fire Departments Learning a New Kind of Fire

One area where lithium battery fires genuinely do present a distinct challenge, separate from the underlying question of how frequently they occur, involves the practical difficulty of extinguishing them once thermal runaway has fully progressed. Fire safety research and reporting from agencies including the United States Fire Administration has documented that electric vehicle fires can take meaningfully longer to fully extinguish than comparable gasoline vehicle fires, and can, in a meaningful proportion of cases, reignite hours after apparently being extinguished, a phenomenon fire safety professionals attribute to the way thermal runaway can propagate slowly between individual cells within a large battery pack, with cells not yet involved in the initial fire event potentially entering runaway themselves hours later as heat gradually conducts through the pack structure, discussed in trade and safety literature examining large scale vehicle fire testing.

This has driven genuine innovation in fire response protocols specifically tailored to lithium battery fires, including the use of specialized fire blankets designed to smother and cool a burning vehicle while limiting the spread of both flame and toxic combustion byproducts, and in some fire department protocols, deliberately submerging a damaged electric vehicle battery pack in water for an extended period, sometimes many hours, specifically to ensure any residual thermal energy within the pack has been fully dissipated before the vehicle is considered safe to move or store, a direct practical response to the reignition risk documented in the research cited above. Large scale, controlled fire testing research comparing battery electric and gasoline vehicle fires directly, conducted in laboratory conditions specifically to generate this kind of comparative data for fire safety professionals, has become an increasingly important area of applied fire science precisely because, as this research explicitly notes, this kind of large scale comparative fire testing data has historically been scarce, given the practical difficulty and cost of safely conducting full scale vehicle fire experiments.

What This Chapter Adds to the Argument

The lithium battery story, considered across both this chapter and chapter five together, offers one of the clearest illustrations in this entire book of the full arc this book has been tracing from its opening pages: a technology capable of storing and releasing a genuinely significant quantity of energy silently, through ordinary controlled operation, that occasionally, under specific triggering conditions, transitions into a dramatic, loud, visually striking failure mode, and that has, through a combination of rigorous root cause investigation following early high profile incidents and sustained, multi layered engineering effort, been made dramatically safer over the roughly fifteen years spanning the earliest widespread consumer lithium battery deployment and the current generation of electric vehicles now statistically demonstrated to be meaningfully safer, on a fire risk basis, than the century old internal combustion technology it is gradually replacing.

This is not a story where the silent hazard was eliminated. Thermal runaway remains a real, physically possible failure mode inherent to lithium battery chemistry, exactly as described in chapter five, and it always will remain possible, for the same fundamental thermodynamic reasons that make lithium chemistry useful for energy storage in the first place. What has changed, dramatically, is the probability of that failure mode actually being triggered in real world use, driven down systematically through exactly the kind of careful, evidence based engineering response that this book will argue, in its concluding chapters, represents the correct and only genuinely effective response to every silent explosion examined throughout these pages: not denial that the hazard exists, and not abandonment of a genuinely useful technology out of excessive caution, but careful, sustained, data driven engineering aimed specifically at understanding the precise mechanism of failure and systematically closing off each pathway by which that failure could occur.

The next chapter turns to industrial settings more broadly, examining reactive metal chemistry, aluminum dust hazards, and the genuine, well documented history of industrial accidents involving materials that, much like the grain dust examined in chapter eight, present little apparent danger in their ordinary bulk form, but become genuinely hazardous once processed, handled, or stored in ways that inadvertently increase their reactive surface area or otherwise shift them further along the spectrum from silent to explosive first established in chapter two.

CHAPTER THIRTEEN

Industrial Silence

At around 7:15 in the evening on February 7, 2008, a series of explosions tore through the Imperial Sugar Company's refinery in Port Wentworth, Georgia, ultimately killing fourteen workers and seriously injuring thirty six others, some with permanent, life altering burns. The material responsible was not a chemical solvent, not a compressed gas, not any substance an untrained observer would immediately associate with explosive danger. It was refined sugar, the same white, granulated substance sitting in kitchen cabinets and coffee shops around the world, transformed into a genuine industrial explosive hazard purely through the same fundamental physical mechanism examined in chapter eight's discussion of grain dust: fine particulate matter, suspended in air within a confined space, ignited by a spark.

The United States Chemical Safety and Hazard Investigation Board's subsequent investigation determined that the initial explosion originated inside an enclosed steel belt conveyor located beneath the facility's sugar storage silos, where recently installed steel cover panels, ironically intended as an improvement to the facility's material handling system, had instead created an enclosed space in which sugar dust could accumulate to explosive concentrations, entirely unnoticed, until a spark, from a source investigators were ultimately unable to definitively identify, triggered the initial blast. That primary explosion, following precisely the mechanism established in chapter eight, dislodged accumulated sugar dust that had been resting undisturbed throughout other areas of the facility for an extended period, feeding a cascade of secondary explosions that ultimately destroyed the facility's packing buildings, its palletizer room, and severely damaged its storage silos and bulk loading areas.

This chapter examines the broader category of industrial hazard that both the Imperial Sugar disaster and the grain elevator explosions examined in chapter eight represent, extending the analysis to materials many people would never intuitively associate with explosive risk, and closing with a careful, deliberately limited examination of one of the more historically significant reactive metal chemistries in industrial history, presented here strictly for its safety relevant and historical significance rather than as any kind of practical guide.

A Hazard Hiding in Ordinary Materials

The United States Chemical Safety Board's landmark 2006 Combustible Dust Hazard Study, examining industrial incidents across a broad range of sectors, documented two hundred and eighty one separate combustible dust incidents occurring between 1980 and 2005 alone, resulting in one hundred and nineteen fatalities and seven hundred and eighteen injuries, according to figures cited in process safety research examining the investigation's findings. What makes this figure particularly striking is the sheer diversity of ordinary, seemingly innocuous materials responsible for these incidents. Beyond the grain dust examined in chapter eight and the sugar dust responsible for the Imperial Sugar disaster, the Chemical Safety Board's broader investigation history includes documented dust explosion incidents involving wood processing dust, various plastics and polymer dusts, pharmaceutical powder dust, and metal dust from processes as varied as machining, grinding, and polishing operations across a wide range of manufacturing sectors.

One particularly instructive case from the Chemical Safety Board's investigation history involves an aluminum dust explosion at the Hayes Lemmerz International manufacturing facility in Huntington, Indiana, investigated and documented in a dedicated Chemical Safety Board report. Aluminum, a metal most people encounter daily in entirely inert, harmless forms, from beverage cans to window frames to aircraft fuselage panels, becomes a genuine explosive hazard once ground into fine dust and allowed to accumulate in an industrial processing environment, for precisely the same surface area principle established in chapter two and reinforced throughout chapter eight's examination of grain dust. This pattern, an ordinary bulk material rendered hazardous purely through particle size reduction, recurs so consistently across industrial dust explosion investigations that safety researchers have organized combustible dust hazards into formal classification systems dating back remarkably far into industrial history, with foundational research on the topic, cited in modern Chemical Safety Board investigation reports, tracing back to a 1924 classification scheme developed by the British researcher R.V. Wheeler, dividing combustible dusts into categories based on their capacity to propagate flame once ignited, a classification framework industrial safety engineers continue to reference and build upon a full century later.

The Regulatory Gap That Persists

Despite the well documented, century old understanding of combustible dust hazards, and despite explicit recommendations from the Chemical Safety Board following its comprehensive 2006 hazard study urging the Occupational Safety and Health Administration to issue a comprehensive federal combustible dust standard applicable across general industry, safety research examining the regulatory landscape as of recent years has noted that such a comprehensive standard remains pending, even as the National Fire Protection Association published its own NFPA 652 Standard on the Fundamentals of Combustible Dust in 2015, a voluntary industry standard developed specifically in response to the continued absence of comprehensive federal regulation, and which has become a widely referenced, though not universally mandatory, framework for industrial combustible dust hazard assessment and mitigation across multiple sectors.

Federal investigators following the Imperial Sugar disaster specifically found, according to figures reported by advocacy organizations and legislative records examining the incident's aftermath, that inspections conducted under the resulting national emphasis program on combustible dust revealed similar underlying engineering and housekeeping failures at facilities well beyond the sugar industry specifically, with some facilities found to have combustible dust accumulation described in official reporting as ankle deep and covering entire rooms, a genuinely striking illustration of exactly the silent, gradual accumulation dynamic established throughout this book: a hazard that builds steadily, invisibly, over an extended period specifically because it does not present the kind of dramatic, attention grabbing signal that would prompt an untrained worker or manager to treat routine industrial dust as the serious, quantifiable explosive hazard it actually represents.

Thermite: A Genuinely Different Category of Reaction

This chapter closes with a careful, deliberately limited discussion of thermite, a reactive metal chemistry that differs meaningfully from every other reaction examined so far in this book, and that deserves inclusion here specifically because of its important, well established legitimate industrial application, discussed here strictly at the level of general chemical principle and historical significance, appropriate for a general audience science book, without any operational or procedural detail.

The thermite reaction is a category of reaction chemists classify as a reduction oxidation, or redox, reaction between a metal and a metal oxide, most classically involving aluminum metal and iron oxide, in which the aluminum, being more chemically reactive, essentially strips oxygen away from the iron oxide, forming aluminum oxide and releasing free elemental iron, along with an extremely large quantity of heat, in a reaction that proceeds at temperatures reported in materials science and welding literature as exceeding two thousand degrees Celsius. Unlike ordinary combustion, this reaction does not require atmospheric oxygen at all, since the oxygen consumed in the reaction comes entirely from the metal oxide reactant itself, a property that makes the reaction capable of proceeding even in environments where atmospheric oxygen is limited or entirely absent.

This specific chemical property, an extremely high temperature, self contained reaction not dependent on ambient atmospheric oxygen, has made a carefully controlled, professionally engineered version of this reaction genuinely useful in a specific, well established industrial application: the welding of railroad rails, a technique that has been used by railway engineering professionals for over a century to join sections of steel rail together into a single, continuous, seamless piece of track, producing a weld with mechanical properties that railway engineers have found to be well suited to the specific stresses of rail infrastructure, a welding technique developed originally in Germany in the late nineteenth century and refined over subsequent decades into the standardized, professionally regulated industrial process still used by railway maintenance crews around the world today.

It is worth being explicit and direct about why this discussion remains at this general, historical level of description rather than proceeding into any further technical or procedural detail. This book's purpose throughout has been to help readers understand the physics and chemistry of energy release phenomena that surround them in ordinary life, industrial settings, and the natural world, not to provide operational guidance for handling genuinely hazardous reactive materials. The professional railway welding application described above is conducted exclusively by trained, certified technicians using purpose built equipment, precise material formulations, and rigorous safety protocols developed and refined over more than a century of professional practice, and readers with a genuine professional or academic interest in this specific application are encouraged to pursue that interest through appropriate professional training and certification channels rather than through a general audience popular science book.

The Pattern That Connects Every Case in This Chapter

Step back from the specific incidents and chemistry examined throughout this chapter, and a single, consistent pattern emerges, one that should feel deeply familiar by this point in the book. Sugar, grain, and aluminum, in their ordinary bulk forms, present no meaningful explosive hazard whatsoever, precisely the same conclusion reached regarding solid iron in chapter three's discussion of rust, and precisely the same conclusion that would apply to nearly any material examined throughout this book in its ordinary, undisturbed, bulk state. Transform that same material, through grinding, processing, or industrial handling, into a form with dramatically increased surface area relative to its mass, allow that transformed material to accumulate silently, gradually, invisibly in a confined industrial space, exactly as documented in the ankle deep dust accumulation findings from federal inspectors following the Imperial Sugar disaster, and introduce a single ignition source, and the result is precisely the kind of catastrophic, loud, structure destroying explosion that opened this chapter.

The engineering and regulatory response to this pattern, comprehensive housekeeping standards, dust collection system design requirements, explosion proof electrical equipment specifications, and the ongoing, if still incomplete, effort toward comprehensive federal regulation specifically targeting this hazard category, represents exactly the kind of systematic, evidence based response to a well understood silent hazard that this book's concluding chapter will argue represents the single most important lesson available across every domain examined throughout these pages. The danger was never mysterious or unpredictable. It was, in every single case examined in this chapter and in chapter eight before it, a direct, physically inevitable consequence of well understood chemistry, allowed to proceed unchecked specifically because the silent, gradual accumulation phase failed to trigger the kind of attention and intervention that a more immediately dramatic, loud hazard would have received far earlier in its development.

The final chapter of this book turns from specific case studies toward a broader synthesis, drawing together the engineering lessons, safety principles, and detection technologies that have emerged across every domain examined in this book, from rust to radioactive decay, from batteries to barns to grain elevators, into a coherent framework for recognizing and responsibly managing silent energy release wherever it occurs, and for understanding where this rapidly evolving field of engineering and materials science may be headed next.

CHAPTER FOURTEEN

Taming the Silent Explosion

The Electric Power Research Institute maintains a database tracking battery energy storage system failures across the growing global fleet of grid scale lithium battery installations, the enormous banks of batteries increasingly deployed by utility companies worldwide to store solar and wind power for use when the sun is not shining or the wind is not blowing. The data this database has accumulated over the past several years tells a genuinely remarkable story, one that offers a fitting note on which to close this book. Between 2018 and 2024, cumulative global battery energy storage deployment grew explosively, from approximately eleven gigawatt hours of installed capacity to more than three hundred gigawatt hours, an expansion of well over twenty five fold in just six years. Over that same period, according to the Electric Power Research Institute's own published analysis of its incident database, the annual number of battery related fire failures remained essentially flat, meaning the failure rate per unit of installed capacity dropped by more than ninety eight percent across those six years.

This is not a story of a dangerous technology being cautiously tolerated despite persistent risk. It is a story of a genuinely significant silent explosion hazard, the same thermal runaway mechanism examined in careful technical detail across chapters five and twelve of this book, being systematically, deliberately, and measurably tamed through sustained engineering effort, informed by exactly the kind of rigorous data collection and root cause analysis this concluding chapter will argue represents the single most important lesson to draw from everything examined throughout this book.

The Common Thread, Restated

Fourteen chapters ago, this book opened by dismantling a popular misconception: that power and danger necessarily announce themselves through noise and flame. Since then, this book has traced that same misconception's dismantling across radioactive decay silently powering spacecraft for half a century, rust quietly consuming the world's infrastructure at a cost measured in hundreds of billions of dollars annually, hay bales cooking themselves into open flame through purely biological and then chemical heating, concrete curing with enough internal fury to threaten one of the largest dams ever built, grain and sugar dust transforming from harmless bulk materials into structure destroying explosives purely through particle size reduction, lithium batteries storing genuinely explosive quantities of energy in devices carried casually in millions of pockets worldwide, phase transitions hiding surprisingly large energy transfers inside the most mundane physical processes imaginable, the human body running a continuous, precisely regulated chemical furnace at every moment of every person's life, and an entire planet radiating forty seven trillion watts of continuous heat from processes occurring far beneath the notice of anyone walking across its surface.

Every single one of these phenomena, examined individually across the preceding chapters, obeys the exact same fundamental principle established in chapter two: energy release does not require sound, and reaction rate, not total energy content, determines whether any given release of stored energy manifests as an imperceptible warmth or a structure destroying catastrophe. What varies from case to case is not the underlying physics, which remains constant throughout, but the specific combination of activation energy, available surface area, confinement, and the presence or absence of self reinforcing feedback mechanisms that determines where any given reaction sits along the spectrum this book established at its outset, running from processes measured in billions of years at one extreme to processes measured in microseconds at the other.

What Engineering Has Learned From Getting This Wrong

If there is a single, unifying lesson available across every case study examined in this book, it is this: the silent phase of any energy release process is not evidence of safety. It is, in every single case examined in these pages, evidence that the danger has not yet been detected, not that the danger does not exist. This distinction, simple to state and yet remarkably easy to overlook in practice, explains why grain elevators burned for decades before comprehensive federal safety regulation finally addressed the underlying dust accumulation hazard, why the Statue of Liberty's iron framework corroded silently for nearly a century before a major restoration project finally revealed the extent of the damage, and why early lithium battery deployments in consumer electronics and commercial aviation experienced genuinely serious, headline generating failures before the multi layered engineering defenses examined in chapters five and twelve were developed and widely deployed.

But this book's final chapter is not intended to end on a note of alarm. Quite the opposite. Every single hazard examined throughout this book, once properly understood, has proven genuinely manageable through careful engineering, and in most cases, that management has become dramatically more effective over time as understanding of the underlying mechanism has deepened. The ninety eight percent reduction in grid battery storage failure rates documented at the opening of this chapter did not happen through luck. It happened because engineers systematically studied every documented failure, identified its specific root cause, whether a manufacturing defect, a design flaw, or an operational practice that increased risk, and closed that specific pathway through targeted engineering intervention, exactly the same fundamental process that transformed grain elevator safety following the catastrophic explosions of the late 1970s, that transformed lithium battery cell manufacturing following the Boeing 787 and Samsung Note 7 incidents, and that transformed hay storage safety practices following generations of accumulated farmer knowledge about moisture content and bale density.

This same disciplined engineering response continues actively today, and it is worth closing this book by looking briefly at where it currently stands. The upcoming 2026 edition of the National Fire Protection Association's Standard 855, governing battery energy storage system safety, will require large scale fire testing specifically designed to demonstrate that modern battery installations can contain a thermal runaway event entirely within a single unit, preventing it from propagating to neighboring units, according to industry research examining current battery storage safety standards, extending the same containment focused engineering philosophy examined in chapter twelve's discussion of electric vehicle battery pack design to the much larger scale of utility grid infrastructure. Battery manufacturers have begun conducting extraordinarily rigorous voluntary testing regimes well beyond current regulatory minimums, including multi day explosion risk assessments combining dispersion testing and deflagration testing, according to industry reporting examining recent large scale battery safety validation programs, a level of proactive, worst case scenario testing that would have been unusual in the earlier, less mature phase of lithium battery deployment examined in chapter five's discussion of the 2013 and 2016 incidents.

The Detection Revolution

Beyond containment and prevention, one of the most significant recent developments across nearly every domain examined in this book involves increasingly sophisticated monitoring and early detection technology, specifically designed to address the core challenge this book has returned to repeatedly: silent hazards are dangerous specifically because they evade ordinary human sensory detection until it is too late to intervene safely.

The wireless hay bale monitoring systems examined in chapter six, capable of tracking internal bale temperature continuously and alerting farmers before dangerous heating crosses the threshold into imminent combustion, represent a direct technological response to precisely this challenge. Modern battery management systems, examined in chapters five and twelve, monitoring individual cell voltage and temperature in real time across battery packs containing thousands of individual cells, represent the same fundamental response applied to a completely different domain. Modern structural health monitoring systems, increasingly deployed on bridges, pipelines, and other critical infrastructure specifically vulnerable to the kind of silent corrosion examined in chapter three, use embedded sensors and increasingly sophisticated data analysis techniques to detect the earliest stages of structural degradation long before it becomes visible to routine visual inspection, addressing exactly the same fundamental problem that allowed the Statue of Liberty's internal corrosion to proceed undetected for the better part of a century.

What unites all of these technological developments, spanning agriculture, consumer electronics, transportation, and civil infrastructure, is a shared underlying philosophy: since these hazards cannot rely on the sound and flame that would naturally draw human attention to a more conventional danger, the appropriate engineering response is to build artificial senses, specifically designed sensors and monitoring systems, capable of detecting the specific physical signatures, temperature, voltage, mechanical strain, chemical composition, that indicate a silent hazard is progressing toward a dangerous threshold, and to make that data available to human decision makers, or in increasingly sophisticated systems, to automated safety systems capable of intervening directly, well before human senses would ever have registered anything was wrong.

A Framework for Thinking About the Next Silent Explosion

This book has focused throughout on well documented, historically established silent energy release phenomena, precisely because grounding every chapter in real, verifiable data was essential to this book's purpose and integrity. But the underlying framework established across these fourteen chapters, energy release without sound, governed by activation energy, reaction rate, surface area, confinement, and feedback mechanisms, is not limited to the specific cases examined in this book. It is a genuinely general framework, applicable to new materials, new technologies, and new engineering challenges that will inevitably emerge in the years and decades following this book's publication.

Anyone encountering a new material, a new energy storage technology, or a new industrial process would do well to ask the questions this book has implicitly asked throughout every chapter. Does this system store genuine chemical, nuclear, or physical energy in a compact form. What determines the rate at which that stored energy can be released, and what conditions or triggers might accelerate that release rate beyond its normal, safe operating range. Does the system involve any self reinforcing feedback mechanism, analogous to the thermal runaway cascade examined in chapters five and twelve, capable of accelerating an initially minor deviation into a full scale, cascading failure. And critically, does the system provide any meaningful sensory warning, to unaided human perception, before it crosses the threshold from silent, controlled operation into dangerous, accelerating energy release, or does it, like nearly every case examined throughout this book, require purpose built monitoring technology specifically to compensate for the fact that human senses evolved to detect fast, loud, visually dramatic threats, and remain largely blind to the slow, quiet, equally real dangers examined throughout these pages.

Where This Leaves Us

The energy was always there. That is, in the end, the simplest possible summary of everything this book has examined across its fourteen chapters. In a rusting bridge support, in a canister of plutonium drifting through interstellar space, in a bale of damp hay stacked in a barn, in the core of a curing dam, in a cloud of grain dust settling silently onto rafters nobody has cleaned in months, in a lithium battery charging quietly on a nightstand, in the ice cube melting in a glass of water, in the sweat cooling your skin after a hard run, in the ground beneath your feet radiating heat from processes that began before complex life existed on Earth, the same fundamental physics has been operating the entire time, releasing genuine, measurable, occasionally enormous quantities of energy, almost entirely without producing a single sound most people would ever think to associate with power or danger.

Understanding this does not require fear. If anything, the appropriate response, and the response this book hopes to have modeled across its fourteen chapters, is a kind of grounded, informed attentiveness, a willingness to look past the absence of noise and flame and ask instead what is actually happening, chemically and physically, in the ordinary objects and processes that surround daily life. That rusting fence post is burning, just very, very slowly. That warm patch of soil might be releasing heat that has been building steadily beneath it for weeks. That phone charging quietly beside your bed contains a genuine, if extremely well managed, capacity for rapid energy release. None of this is cause for alarm. It is, instead, an invitation to see the world a little more clearly, to recognize that silence has never been the same thing as safety, and that some of the most powerful, most consequential processes on this planet, and beyond it, have been happening all along, patiently, invisibly, in complete and total silence, waiting only for someone to understand exactly what they were looking at.

EPILOGUE

A Final Word

Somewhere behind you, right now, wherever you happen to be reading this final page, something is very likely rusting. A hinge, a nail, a fence post, a car body panel, a length of rebar buried inside a concrete wall. It has probably been rusting for longer than you realize, and it will probably continue rusting for years after you have forgotten this book entirely. That quiet, ongoing process is, chemically, a fire. It has simply never once made a sound.

That is the thought this book has been built around, examined from as many honest, well documented angles as fourteen chapters could hold. Not a trick of language, not a rhetorical flourish, but a genuine, physically accurate description of what is actually happening in that rusting metal, in that curing concrete, in that quietly decaying radioactive isotope, in that battery sitting on your nightstand, in your own body at this exact moment, converting the last meal you ate into the warmth keeping you comfortable as you read these words.

If this book has done its job, you will not walk away from it afraid of rust, or batteries, or hay bales, or your own metabolism. Fear was never the goal, and fear, in any case, tends to be a poor substitute for understanding. What this book hopes you carry forward instead is a slightly sharper way of seeing the ordinary world, a willingness to ask, when something is generating heat, or slowly changing, or quietly building toward some threshold, what is actually happening here, chemically and physically, rather than assuming that silence and stillness mean nothing of consequence is underway.

The engineers, chemists, physicists, and safety researchers whose work populates every chapter of this book did not solve the problems examined here by being afraid of silent energy release. They solved them by taking that silence seriously, by refusing to assume that the absence of noise meant the absence of danger, and by doing the patient, unglamorous work of measurement, testing, and root cause analysis that eventually turned mysterious, catastrophic failures into well understood, manageable engineering challenges. That is not a dramatic story in the way a bomb going off is dramatic. But it may be, in its own quiet way, a more genuinely impressive one, and it is the story this book has tried, throughout every chapter, to tell honestly and completely.

The world is considerably more energetic than it appears to be at first glance. Most of that energy, most of the time, is moving through its transformations quietly, safely, and usefully, exactly as engineers and natural processes alike have arranged for it to do. Understanding where that energy is, and how it behaves, is not a cause for anxiety. It is, if anything, one of the more genuinely satisfying kinds of knowledge available, the kind that changes how an ordinary walk past a construction site, a scrapyard, or a farm feels, permanently, in a small but real way.

Thank you for reading.

ABOUT THE AUTHOR

Anton Kesting writes about the hidden physics and chemistry of everyday life, translating technical and scientific research into accessible narrative nonfiction for general readers. His work focuses on finding genuine, well documented scientific stories in ordinary, overlooked corners of the physical world, from industrial engineering and materials science to the quiet biochemistry of the human body. He lives with an ongoing curiosity about the reactions happening quietly, all around us, all the time.

A NOTE ON SOURCES

Every specific figure, historical incident, and scientific claim presented in this book is drawn from published research, government investigation reports, peer reviewed scientific literature, and reporting from established scientific and news organizations. Where official investigation findings exist, including reports from the National Transportation Safety Board, the United States Chemical Safety and Hazard Investigation Board, the Occupational Safety and Health Administration, and NASA, this book has drawn directly on those findings. Readers interested in exploring any topic covered in this book in greater depth are encouraged to consult the original agency reports and peer reviewed research referenced throughout these chapters, many of which are freely available through the issuing agencies and academic publishers.

www.ingramcontent.com/pod-product-compliance
Lightning Source LLC
LaVergne TN
LVHW100144061126
851857LV00010B/231
* 9 7 9 8 2 3 5 5 5 8 5 6 4 *